T0182132

Fifty Materials That Make the World

Ian Baker

Fifty Materials That Make the World

with a Foreword by Michael F. Ashby

 Springer

Ian Baker
Thayer School of Engineering
Dartmouth College
Hanover, NH, USA

ISBN 978-3-030-08780-7   ISBN 978-3-319-78766-4   (eBook)
https://doi.org/10.1007/978-3-319-78766-4

Cover illustration: The first figure is a hardwood Ulmus (Elm) provided by Sara Essert. The second
figure is of a wool fiber provided by Zhangwei Wang

Printed on acid-free paper

This Springer imprint is published by the registered company Springer International Publishing AG
part of Springer Nature.
The registered company address is: Gewerbestrasse 11, 6330 Cham, Switzerland

This book is dedicated to my mother and father, Margaret Noreen Baker and John Henry Baker.

Foreword for Fifty Materials That Make the World

We live in a world of materials; it is materials that give substance to everything we see and touch. Our species – *Homo sapiens* – differs from others most significantly, perhaps, in its ability to *design* – to make things out of materials – and in the ability to see more in an object than merely its external form. Designed objects, symbolic as well as utilitarian, predate any recorded language – they provide the earliest evidence of a cultural society and of symbolic reasoning. Objects have meaning, carry associations, and become symbols for more abstract ideas. All these are inseparable from the materials of which the objects themselves are made.

Materials have engineering associations: the strength of steel, the conductivity of copper, the temperature tolerance of tungsten. Beyond that, they have aesthetic associations: the warmth of wood (the paneled parlor), the perkiness of plastic (the material of Pop art), the chunkiness of concrete (roadblocks, sea defences), and the aura of gold (the embodiment of bling). And more subtly, materials become metaphors: an iron will, a leaden sky, a glassy look, a silver lining, a glided life. Materials, in these many senses, *are* our world.

The essay, a literary form popular in the eighteenth century, is a short composition based on a single subject, often colored by the personality of the author. We live today in the era of over-long novels and even longer textbooks (Shackelford, *Materials Science for Engineers* 6th ed.: 878 pages; Callister and Rethwisch, *Materials Science and Engineering* 9th ed.: 936 pages, and, I confess, my own *Materials Selection in Mechanical Design*: 646 pages), with little that is personal in their style. In stark contrast, Ian Baker has harnessed the charm and engagement of the essay form in constructing this assembly of portraits of *Fifty Materials That Make the World*.

Here is a taste of what you will find, each short chapter an amalgam of solid scientific fact and pleasing anecdote.

ABS, the material of computer keyboards, telephones, and childrens' toys, emerges as a star among plastics, stronger than the strongest polyethylene and tough enough to survive anything an active 6-year-old can devise to destroy it.

Concrete, used since 6,500 BC and, much later on, the glue of Roman domes and aqueducts, is used today in greater quantities than any other material. And for good reasons: it is cheap, strong, durable, and uses widely available raw materials. But – there is always a "but" – the sheer quantity makes it one of the largest sources of carbon emissions to atmosphere, something that scientists today seek to overcome.

Diamonds, we learn, are durable, but not, perhaps "for ever." The stable form of carbon at atmospheric pressure is not diamond, so – with kinetics too slow to be perceived on a human time-scale, that sparkler is slowly reverting to a much less precious allotrope – graphite.

Paper, a disruptive technology, purveyor of information in printed form and much more for almost 2000 years, was to have met its demise by now as the paper-less society displaced it. Not so: paper production has tripled since 1978 when that prediction was made. Paper is survivor.

These 50 chapters stimulate an appetite for materials: they are accessible, tasty, digestible, leaving you wanting more. And more is provided in the form of copious links and references. Dive in.

Cambridge, England   Michael F. Ashby
May 2018

Acknowledgments

I would like to acknowledge many people who have provided fascinating images for this book. They are individually acknowledged after the relevant image.

Finally, in writing this book, I would like to acknowledge the invaluable assistance of Stephanie Turner. Stephanie not only typed much of the text, but she also drew several of the figures. More importantly, she read each chapter as a non-specialist and gave me invaluable feedback. She helped to make the text far more intelligible, and any remaining incoherent sections are purely my own work.

Contents

About the Author

Ian Baker obtained his B.A. and D. Phil. in Metallurgy and Science of Materials from the University of Oxford. He then joined the Faculty of the Thayer School of Engineering at Dartmouth College, where he is currently the Sherman Fairchild Professor of Engineering and Senior Associate Dean for Academic Affairs. He is the Editor-in-Chief of the journal *Materials Characterization*.

Introduction

The aim of this book is both to introduce the nonspecialist to the excitement of Materials Science and to present to Materials Scientists some materials and history of materials that they may not know. While the book is about materials, their structure and properties, it is also a history book. However, the history is not linear since the history associated with the discovery and development of each material is presented in separate chapters.

The seed for this book was sown by a 2010 BBC series of podcasts entitled "A History of the World in 100 Objects" by Neil MacGregor when he was director of the British Museum. Each of the 100 15-min podcasts was about an object that resides in the British Museum, which he described and positioned in a wider historical landscape. Two other books that influenced my thoughts covering a select number of objects and their place in history were *Fifty Machines That Changed the Course of History* and *Fifty Minerals That Changed the Course of History* both of which are by Eric Chaline.

I had also read a number of books that aimed to convey the excitement of Materials Science to the nonspecialist. The first, *The New Science of Strong Materials or Why You Don't Fall Through the Floor* by J.E. Gordon, was so influential on me while at Grammar School that I embarked on studying Materials Science at university – and ever since. Two books that I found particularly interesting were *The Substance of Civilization* by now-retired Cornell Professor of Materials Science Stephen Sass and *Stuff Matters: The Strange Stories of the Marvellous Materials That Shape Our Man-Made World* by Professor Mark Miodownik of University College, London. I know Steve because we both worked on the intermetallic compounds. His book was based on a lecture course that he gave at Cornell, and it was subsequently the basis of a lecture course at the Thayer School of Engineering, Dartmouth College, that has been taught for many years by Professor Ron Lasky. Although Mark and I have never met, both he and I attended St. Catherine's College, University of Oxford, for our B.A. and D. Phil., but years apart. And to indicate what a small, interconnected world it is, his father offered me a position as lecturer at the University of Surrey over 30 years ago. It is worth noting that these three books are both of interest to the Materials Scientist and the non-Materials Scientist.

Putting the two ideas of conveying the excitement of studying materials to the non-Materials Scientist and centering this on a limited number of materials led to this book. Apart from focusing on a limited number of materials, I have also, at the possible expense of losing many potential readers, made the book more quantitative than those cited above. As a scientist, I feel that I cannot simply say one material is stronger than another or a material has better conductivity. I need to say by how much. Thus, one has to provide values for these properties. For the nonscientist, this may seem intimidating, but to quote from the cover of the Hitchhiker's Guide to the Galaxy "DON'T PANIC." All you have to do is look at the numbers and do not worry about the units accompanying the numbers. For example, when stating Alloy A at 500 MPa is significantly stronger than Alloy B at 300 MPa, you do not need to worry that Pascals (Pa) is a unit or measure of pressure or stress or that the M stands for "Mega," that is a million, all you have to do is compare the "500" with "300" to understand the difference in strength since the units are the same.

I have largely used the *Système International d'Unités* or SI units (meters, kilograms, etc.) rather than Imperial of English units throughout. Thus, I have used the metric "tonne" rather than "ton." The British ton and the metric tonne are quite close, whereas the tonne is 1.1 US tons. Most materials are sold on world markets in tonnes. Again, one need not worry about the units since the same ones are used when comparing properties: just look at the numerical values. However, I have used degrees centigrade (°C) throughout rather than the SI unit of Kelvin (K) since few people who are not scientists can relate to the latter measure of temperature. Another deviation from using SI units was for some precious materials. Platinum and some other materials are sold in troy ounces, which are about 1.1 avoirdupois ounces, and so I have quoted some prices in troy ounces. Similarly, for diamonds I quoted weights in both SI units and carats since the latter is used when weighing diamonds and other precious stones.

Choosing the 50 materials, which is clearly an arbitrary number, was not straightforward. I started by polling some Materials Scientists at the Thayer School of Engineering, Dartmouth College, on which are the ten most important materials. While there was not agreement on all ten materials, there was a large amount of overlap. I think I would obtain a similar answer if I asked many Materials Scientists. Everyone would agree on steel, aluminum alloys, stone, bronze, polythene, and many others as being important. Coming up with 30 or so important materials was fairly straightforward. However, in deciding which materials should fill out the 50 was more difficult and, to some extent, is a matter of opinion depending on what criteria are used. Some of the materials may seem odd to include such as Gutta Percha and Bakelite since these are not used greatly these days. However, at the time of their introduction they were groundbreaking materials and had a large impact. One could make the same point about bronze. Bronze is still used, but this is no longer the Bronze Age when bronze was *the* material. In fact, in writing this book, I replaced some of the chapters from the first to second draft as it became clear some materials were likely more important than others.

Before leaving the choice of which materials to cover in 50 chapters, I have a *mea culpa*. Some chapters cover more than simply the material stated in the chapter title. For example, the chapter on rare earth (RE) magnets starts by discussing the first engineered permanent magnets, AlNiCo alloys, whose market share is now quite small and declining, and the materials that displaced them, ferrites, before discussing RE magnets, which are now the most valuable part of the market.

To conclude, I should point out that I am not an expert on all the materials covered in this book. What I have only attempted to cover are the main points about a material. For the interested reader, there are whole books – in some cases many books – devoted to the materials in some chapters. I should also note that I am not a historian and so I am indebted to the people's work that I have quoted. Any errors in interpretation of that work are my own.

Chapter 1
ABS Plastics

If you Google "ABS", you'll likely get "anti lock brakes" or a reference to stomach muscles. ABS is also the acronym for Acrylonitrile Butadiene Styrene, the most popular engineering polymer - an engineering polymer is one that is used because of its mechanical properties. You have undoubtedly pounded on this material since one of its many applications is for computer keyboards. ABS plastic is an amorphous (lacks the long range order associated with crystals) thermoplastic (one that can repeatedly re-melted) with a glass transition temperature (when it gets significantly softer) of 105 °C and, thus, is easy to manufacture into products by extrusion or injection molding at relatively low temperatures of 204–238 °C. It is also easy to machine.

The Borg-Warner Corporation, Auburn Hills, Michigan patented ABS in 1948 and introduced it commercially in 1954.[1] ABS plastic [2] is quite different to the most commonly-used polymer, polyethylene, which consists of a single monomer ($-CH_2-CH_2-$) repeated over and over again. ABS plastic is made though an emulsion process from varying amounts of the monomers acrylonitrile (15–35%), 1,3 butadiene (5–30%) and styrene (40–60%), see Fig. 1.1, and, thus is referred to as a terpolymer. The range of compositions means that the formula for ABS plastic is not fixed but can be written $(C_8H_8)_x(C_4H_6)_y(C_3H_3N)_z$, where x, y and z represent the proportions of acrylonitrile, butadiene and styrene. This range of compositions also means that the density of ABS plastic varies accordingly from 900–1530 kg/m³, that is from less dense than water (1000 kg/m³) to heavier than water.

The mechanical properties of an ABS plastic also depend on the proportions of the three monomers. Microstructurally, ABS plastics consist of sub-micron particles of long chains of rubbery butadiene, which provide toughness, in an amorphous matrix consisting of shorter chains of the styrene-acronitrile copolymer in which the polar nitrile groups join the chains together. The latter also provide a shiny

---

[1] Acrylonitrile-butadiene-styrene copolymer (ABS), chemical compound, Britannica.com.pdf.

[2] http://www.bpf.co.uk/plastipedia/plastics_history/Default.aspx

© Springer International Publishing AG, part of Springer Nature 2018
I. Baker, *Fifty Materials That Make the World*,
https://doi.org/10.1007/978-3-319-78766-4_1

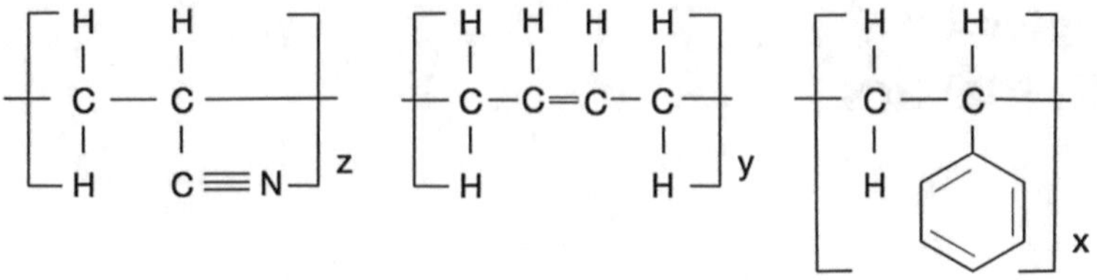

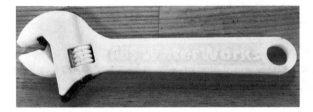

A: polyacrylonitrile   B: polybutadiene   S: polystyrene

Fig. 1.1 The three monomers that are used in different proportions to make ABS plastic.

Fig. 1.2 A fully working adjustable wrench that was produced from ABS without any need for assembly using a rapid prototyping machine. The black pieces are support material that was used during the processing

impervious finish.[3] This structure means that ABS plastics have a tensile strength (37–52 MPa [4]) that is higher than that of the strongest polyethylene, ultra high molecular weight polyethylene, at around 30 MPa. They also have a good elastic modulus (resistance to stretching) of 2.3 GPa (for a polymer), good elongation to failure of up to 70%, and excellent durability from −20–80 °C [1]. ABS belongs to a class of materials called rubber-toughened thermoplastics in which the rubber particles dissipate the energy at growing crack tips. They are, thus, much tougher than other plastics. ABS can be further strengthened by the inclusion of glass fibers.

ABS plastic is used in a wide variety of applications including various kitchen appliances, the inside of refrigerators, vacuum cleaner housings, telephones, toys (such as Lego), drain pipes, power tool housings, electrical socket covers, musical instruments, luggage, lawn and garden equipment, automotive parts (such as bumpers and trim), and electronic or electrical casings.[5] ABS plastic is also used in some rapid prototyping machines, see Figs. 1.2 and 1.3. It ranges in color from white to ivory and is opaque, but can be colored with dyes for various applications and it can be sanded and painted. Since ABS plastic can be degraded by ultraviolet light, additives are sometimes used to retard this degradation.

---

[3] Helbig, M and Seelig, T. "Multiscale modeling of deformation and failure in ABS-materials,13. Problemseminar, Deformation und Bruchverhalten von Kunststoffen" Merseburg, Juni 2011, 011_MH_ThS_Multiscale.pdf.

[4] http://www.makeitfrom.com/material-properties/Acrylonitrile-Butadiene-Styrene-ABS

[5] PlasticsEurope - Acrylonitrile-Butadiene 1-Styrene (ABS) - PlasticsEurope.pdf.

Fig. 1.3 Three nesting spheres made from ABS that were produced one inside the other using rapid prototyping

ABS plastic has replaced polystyrene in many applications because of its superior strength and toughness, and better finish - even though it is twice the cost at about $1.50 per kg.[6] It also has excellent resistance to some acids and alkalis. Since ABS polymers are thermoplastic they can relatively easily be recycled.

ABS is expected to remain the leading engineering plastic [7] with a market size of $22.3 billion for 2015, which is expected to grow at 6% per year.[8]

Reference

1. Smith, W. F. (1990). *Principles of materials science and engineering* (2nd ed.). New York: McGraw-Hill Publishing Company. ISBN: 0-07-059169-5.

---

[6] Everything You Need to Know About ABS Plastic.pdf.

[7] http://www.marketsandmarkets.com/PressReleases/engineering-plastic.asp

[8] https://www.gminsights.com/industry-analysis/acrylonitrile-butadiene-styrene-ABS-market

Chapter 2
Aluminium/Aluminum

This is not like the lyrics of the George and Ira Gershwin song "Let's Call the Whole Thing Off" sung by Fred Astaire and Ginger Rogers in the 1937 film "Shall We Dance", of "you say tomatoes (to-may-toes), I say tomatoes (to-mah-toes)".[1] Americans and British pronounce the name of this element differently because they spell it differently. The metal was named aluminum even before it was first isolated by the British chemist Humphrey Davy (1778–1829) in 1807. The name was later changed to aluminium to bring its spelling more in line with other metallic elements that end in "ium", a name that was adopted by the International Union of Pure and Applied Chemistry. Oddly, in 1925 the American Chemical Society decided to revert to the original name aluminum, which is now the usage in America, while elsewhere in the English-speaking world it is still aluminium. Whichever spelling you use, the name is derived from Alum, the common name of hydrated potassium aluminum sulfate.

Aluminum is ubiquitous in our daily lives. It is the third most abundant element on earth and the most abundant metallic element in the Earth's crust comprising a little over 8%. It is a face-centered cubic metal (see Fig. 2.1) that finds use in almost pure form in beverage cans and cooking foil. In highly-alloyed form it is used in such high strength applications as aircraft wings and bodies, automobiles, and engines. Since the introduction of the first commercial airliner, the ill-fated British de Havilland Comet, in 1952 the body and wings of commercial aircraft have been made largely of aluminum. However, aluminum is now losing its dominance and the body and wings of Boeing's 787 Dreamliner's are made largely of a carbon-fiber-reinforced polymer. Whilst widespread today, aluminum, is the most recently isolated of the commonly-used metals. It was produced for the first time in 1825 by the Danish scientist Hans Christian Oersted (1777–1851), who reacted anhydrous aluminum chloride with an alloy of potassium and mercury, and then heated the resultant material under reduced pressure.

---

[1] https://www.letssingit.com/fred-astaire-feat.-ginger-rogers-lyrics-let-s-call-the-whole-thing-off-dksxl8n

© Springer International Publishing AG, part of Springer Nature 2018
I. Baker, *Fifty Materials That Make the World*,
https://doi.org/10.1007/978-3-319-78766-4_2

Fig. 2.1 The face centered cubic structure adopted by aluminum and a number of other metals such as copper and nickel. The figure shows the atoms as they pack together touching each other and filling roughly 74% of space. The length of the side of the cube, the lattice parameter, is 0.405 nm for aluminum

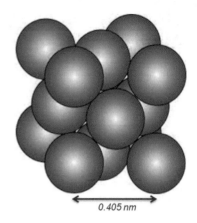

0.405 nm

The late discovery of aluminum compared to other metals and the difficulty of its extraction are because aluminum bonds very strongly to oxygen, and thus, does not occur naturally in its metallic form. It is extracted from the ore Bauxite, a mixture of oxides, hydroxides and a few other materials such as clay. It was not until the simultaneous but independent invention of what is now called the Hall-Héroult process in 1886 through which aluminum is extracted by electrolysis from aluminum oxide dissolved in molten cryolite (Na_3AlF_6) by a Frenchman, Paul Héroult (1863–1914) and an American, Charles Martin Hall (1863–1914) that aluminum became affordable for everyday applications [1].

Before the electrolytic processing route, aluminum was so expensive that, famously, cutlery was produced for Louis Napoleon Bonaparte (1808–1873), the only president of the short-lived French Second Republic (1848–1851) and Emperor of the Second French Empire (1852–1870) as Napoleon III, for use by only his most honored guests [2]. In 1884, it was installed as a lightening conductor on the 23 cm tall, 2.8 kg apex of the Washington monument [2], a 169 m high marble obelisk. Now at $1.93/kg, the price of aluminum is three times the price of rolled mild steel ($0.65/kg) by weight, but is actually a similar price per unit volume. This is a sharp drop from the $200/kg in 1856 - the price fell to $0.65/kg in 1888 after Charles Hall started the Pittsburgh Reduction Company, which later became the Aluminum Company of America or Alcoa.

Several properties of aluminum make it very useful. First, it is very resistant to oxidation since the oxide that forms alumina, Al_2O_3, has the same density as aluminum and a thin protective layer readily and rapidly forms at room temperature due to it high heat of formation of -1670 kJ/mol. By comparison, the heat of formation of copper oxide is only -155 kJ/mol, which is why one sees shiny pieces of copper but not aluminum. The good news is this rapid oxide formation is very protective and so aluminium does not need to be protected by coating or painting to prevent oxidation. Second, aluminum has both excellent thermal conductivity (237 W/m·K at 25 °C) and electrical conductivity (37×10^6 Siemens.m^{-1} at 20 °C). Cookware and wiring make use of these properties along with the low cost and oxidation resistance. The property that is really advantageous in many applications is its low den-

Fig. 2.2 Transmission electron micrograph showing Mg$_2$Si precipitates (arrowed) aligned lying on specific planes in aluminum alloy 6005A-T6. (Courtesy of Aude Simar)

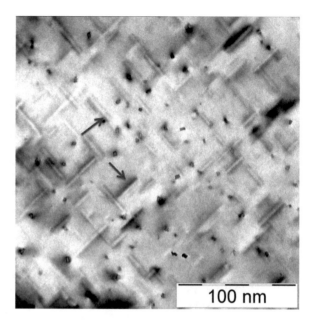

100 nm

sity (2712 kg/m^3), which at 2.7 times the density of water is only about one third that of steel (7850 kg/m^3). This low density makes it useful for aircraft production. Unfortunately, aluminium, like all unalloyed (pure) metals is quite soft with a yield strength of only 7–11 MPa. In 1903, Alfred Wilm (1869–1937), who worked at Dürener Metallwerke Aktien Gesellschaft, found that by adding just 4% copper to aluminium the strength could be greatly improved, and in the process he discovered so-called age-hardening whereby a alloy can be heat-treated and can get stronger at longer times. The first aircraft alloy, a stronger version of this aluminum-copper alloy, containing an additional 1% manganese and 1% magnesium, was developed in 1909. Since the company was in Düren, Germany, the improved alloy was given the name Duralumin.

When Duralumin is heated to 500–510 °C all the alloying elements are in solution in the aluminium, and the alloy after quenching to room temperature is relatively soft and ductile and can be easily rolled, extruded, drawn or forged. The alloy can be naturally aged at room temperature or artificially aged at 190 °C, which accelerates the aging, so that very fine particles of CuAl$_2$ and Mg$_2$Si precipitate in the alloy. These make it much stronger with a yield strength of ~450 MPa resulting in a strength-to-weight ratio, often called specific strength (strength at failure divided by the density) of 240 kN · m/kg comparable to some steels. Most modern aluminum alloys utilize precipitates for strengthening, often Mg$_2$Si, see Figs. 2.2 and 2.3. Unfortunately, the alloying additions make the aluminum more susceptible to corrosion, but Duralumin can be clad with aluminum to protect it, resulting in a product called Alclad. Unfortunately, this is also more expensive than aluminum at ~$4/kg.

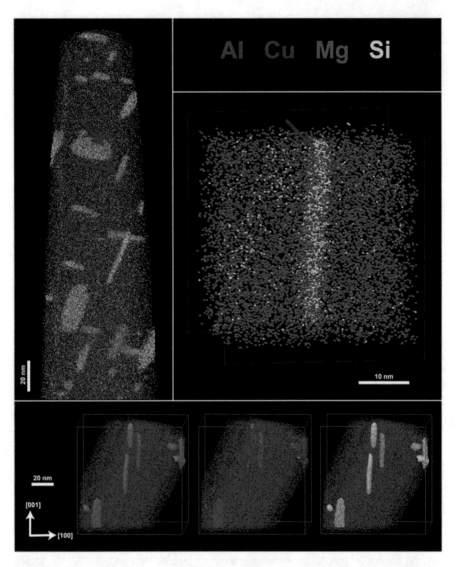

Fig. 2.3 "Atom probe microscopy (APM) of an experimental 6xxx-series Al-Mg-Si-Cu alloy after age hardening. These atom maps provide tomographic images that are a quantitative depiction of the local composition and the relative proximity of each of the matrix (Al) and the various solute atom species, with respect to each other. Mg-Si-Cu rich precipitates are observed. (**a**) A Si-Mg-Cu atom map, providing a good overview image of the solute-rich precipitates, which are seen to possess a narrow plate-shaped morphology (arrowed). (**b**) An Al-Si-Cu atom map providing a high resolution view of the precipitate-matrix interface (arrowed). (**c**) A series of iso-concentration surfaces that bound regions of a particular concentration level for Mg, Cu and Si in the precipitates. These various forms of APM data can serve as the starting point for materials scientists to investigate precipitate morphology, composition, size, spacing, number density, and the tendency for pre-precipitate solute atom clustering to occur in the matrix." (Courtesy of Suqin Zhu, Alec Day and Simon Ringer)

Fig. 2.4 Aluminium is omnipresent as a beverage container

Nowadays, there are several series of aluminum alloys from the 1000 series to the 8000 series that have different alloying additions and, thus, different properties and costs, which are used for different applications. Aluminum use continues to grow: production increased from 2.3 tonnes in 1873 to 6600 tonnes in 1890 to 44 million tonnes in 2014. With increasing emphasis on energy efficiency, aluminum's use is likely to grow in applications such as car bodies, such as Tesla automobiles, which utilize its low density and corrosion resistance. One of the ever-present uses of aluminum is in beverage cans, see Fig. 2.4.

Finally, if you are going to recycle, aluminum is a must. It only takes 5–6% of the energy to melt aluminum at 660 °C compared to winning it from its ore. Thus, aluminum recycling is extensive and every aluminum product that you use is around 50% recycled material, a proportion that is expected to grow.

In 2016, World production of aluminum was nearly 58 million tonnes, with an astounding 31 million tonnes produced in China – the next highest producer was Russia at only 3.6 million tonnes. By comparison, the production of steel, the most produced metal, was 1630 million tonnes. On February 14th, 2018 aluminum was around $2177/tonne, which is about halfway between its low of the last twenty five years of $1146/tonne on March 8th, 1999 and its high of $3025 on July 8th, 2008.[2]

References

1. Chaline, E. (2012). *Fifty minerals that changed the course of history*. Buffalo: Firefly Books. ISBN: 13: 978-1-55407-984-1.
2. Amato, I. (1998). *Stuff: The materials the world is made of*. New York: Avon Books, Inc. ISBN-10: 0380731533.

[2] http://www.macrotrends.net/2539/aluminum-prices-historical-chart-data

Chapter 3
Asphalt

If asked what is one of the most recycled materials, you might not expect the answer to be asphalt concrete.[1,2] Asphalt concrete roads, that is roads made from a mixture of around 5% asphalt with an aggregate that contains stone, sand and gravel, date to only 1903 when British inventor Edgar Purnell Hooley (1860–1942) obtained a patent for tarmac in which he premixed asphalt and aggregate that was laid on the road and steamrolled. Hooley subsequently started the Tar Macadam Syndicate Ltd. The first tarmac road was Radcliffe Road in Nottingham, England. Tarmac built on earlier work by Scotsman John Loudon McAdam (1756–1836) who pioneered the use of macadam roads in the 1820s [1]. Unfortunately, these early roads were less than optimum and were prone to generate dust and to rutting. Thus, around 1834, John Henry Cassel improved road building by starting with a tar layer, adding a layer of McAdam's material and finishing with a mixture of tar and stone [2].[3]

Although asphalt roads are quite recent, the use of asphalt, which is also often called bitumen, dates back to at least the 5th millennium B.C. when it was used in the Indus valley for sealing baskets. Later, it was used by the Sumerians, in modern-day Iraq, as a mortar, a glue and for waterproofing, while the Ancient Egyptians even used it in the embalming process for mummies. The Byzantine navy likely used asphalt as one of the ingredients in the Greek Fire that they siphoned onto the Arab Fleets that attacked Constantinople in both 677 A.D. and 717–718 A.D. to defeat them [3].

Asphalt is sticky, black, highly-viscous liquid that is found naturally, but these days is mostly produced from oil. The World's largest deposit of asphalt is the oil sands, which are unconsolidated sandstone, in Alberta, Canada. That particular deposit is being mined and turned into synthetic oil at enormous costs to the environment as well as producing very expensive oil. Asphalt is formed from organisms

---

[1] http://www.asphaltpavement.org/recycling

[2] http://asphaltrecycling.com/display.php?cnt_id=24

[3] http://www.bbc.co.uk/nottingham/content/articles/2009/07/03/edgar_hooley_tarmac_feature.shtml

© Springer International Publishing AG, part of Springer Nature 2018
I. Baker, *Fifty Materials That Make the World*,
https://doi.org/10.1007/978-3-319-78766-4_3

Fig. 3.1 One advantage of asphalt roads is that cracks can be repaired with liquid asphalt

that died and ended up at the bottom of a lake or ocean and were subsequently subjected to large pressures and temperatures over 50 °C deep within the earth. The resulting asphalt is a complex mixture of many organic chemicals with the largest components being partially hydrogenated aromatic compounds, high molecular weight phenols, and carboxylic acids.[4,5]

Most of asphalt's use throughout history prior to road building was for sealing and as a glue or mortar. However, in 1826/7 the French scientist Joseph Nicéphore Niépce (1765–1833) used an asphalt coating on a pewter plate to produce the first photograph of a nature scene. The asphalt when exposed to sunlight hardens.[6] Thus, areas that receive more sunlight harden more. After an exposure of 6–7 h, the softer asphalt can be washed off to produce an image. Obviously, not a process useful for action photography. This property of asphalt enabled its use as a photoresist for making printing plates, a technique used from around 1850 to around 1920.

Nowadays, asphalt is largely used for roads, see Fig. 3.1. In the U.S.A., 85% is used for asphalt concrete roads, which are typically 5% asphalt and 95% aggregate. It is used for road repair as Asphalt Emulsion, which is up to 70% asphalt. This is sprayed onto roads followed by adding a layer of crushed stone. For housing applications, Mastic Asphalt, which consists of 7–10% asphalt, is used for waterproofing flat roofs and it is used to make roofing shingles.

---

[4] http://www.e-asphalt.com/ingles/composition.htm

[5] https://pubs.acs.org/cen/whatstuff/stuff/7747scit6.html

[6] https://www.britannica.com/biography/Nicephore-Niepce

The future may involve the modification of asphalt with polymers such as styrene-butadiene (synthetic rubber), reclaimed tire rubber, polyethylene, and atactic polypropylene, to improve the properties of asphalt. Polymers can increase the useful temperature range considerably and increase the useful lifetimes of roofs and roads by up to a factor of ten [4]. The downside is that polymer additions can increase the cost of the asphalt by 60–100%. Thus, the use of polymer-modified asphalt depends strongly on price.

References

1. van Dulken, S. (2000). *Inventing the 20th century: 100 inventions that shaped the world. From the airplane to the zipper*. New York: New York University Press. ISBN: 0-8147-8808-4.
2. (1848). *Repertory of patent inventions and other discoveries and improvements in arts, manufactures and agriculture* (Vol. 12). London: Alexander Macintosh.
3. Chaline, E. (2012). *Fifty minerals that changed the course of history*. Buffalo, New York: Firefly Books Ltd..
4. Becker, Y., Méndez, M. P., & Rodríguez, Y. (2001). Polymer modified asphalt. *Vision Tecnologica, 9*, 39–50.

Chapter 4
Bakelite

Invented in 1907 in Yonkers, New York, phenolformaldehyde was the first thermo-setting polymer, that is, a polymer that once it has set can't be remelted. The polymer was called Bakelite, or more properly Baekelite after its inventor the American chemist Leo Henricus Arthur Baekeland (1863–1944).[1] One of its first uses may have been for the knob on a gear lever in a Rolls Royce automobile [1]. Bakelite provided the long-sought solution to replace ivory for billiard balls and is now also used for bowling balls [2].

Bakelite is produced by the condensation reaction – water is the condensate molecule - between phenol and methanal (formaldehyde) in the presence of a catalyst. The resulting product is a brown, moldable, insulating, heat-resistant, water-resistant polymer with a density of 1300 kg/m^3, that is a bit denser than water at 1000 kg/m^3,[2] see Fig. 4.1. The extensive cross-linking produces a three-dimensional polymer that is both hard and brittle. Bakelite can be strengthened by the addition of cellulose from wood pulp [3].

Bakelite's uses stemmed from its high electrical resistivity, heat-resistance, scratch resistance, resistance to solvents, and its low cost for producing parts in quantity. Thus, it was used for radio and telephone casings, kitchenware, jewelry, children's toys, firearms and miscellaneous parts like a distributor rotor. Bakelite products were molded from powder or partially-cured slugs under temperature and pressure, and the resulting product had a smooth, shiny surface that needed little in the way of finishing. Bakelite resin can also be applied to layers of paper, glass or many fabrics that under heat and pressure produce Bakelite phenolic sheets of a wide variety of properties that can be used for a range of applications. In the 1920s and 1930s Bakelite was an exciting new material with new properties that was considered the material expressing the sentiment of the age.

---

[1] The Bakelizer, National Museum of American History Smithsonian Institution, November 9, 1993, The American Chemical Society.

[2] Commercially Important Condensation Polymers | Tutorvista.com

© Springer International Publishing AG, part of Springer Nature 2018
I. Baker, *Fifty Materials That Make the World*,
https://doi.org/10.1007/978-3-319-78766-4_4

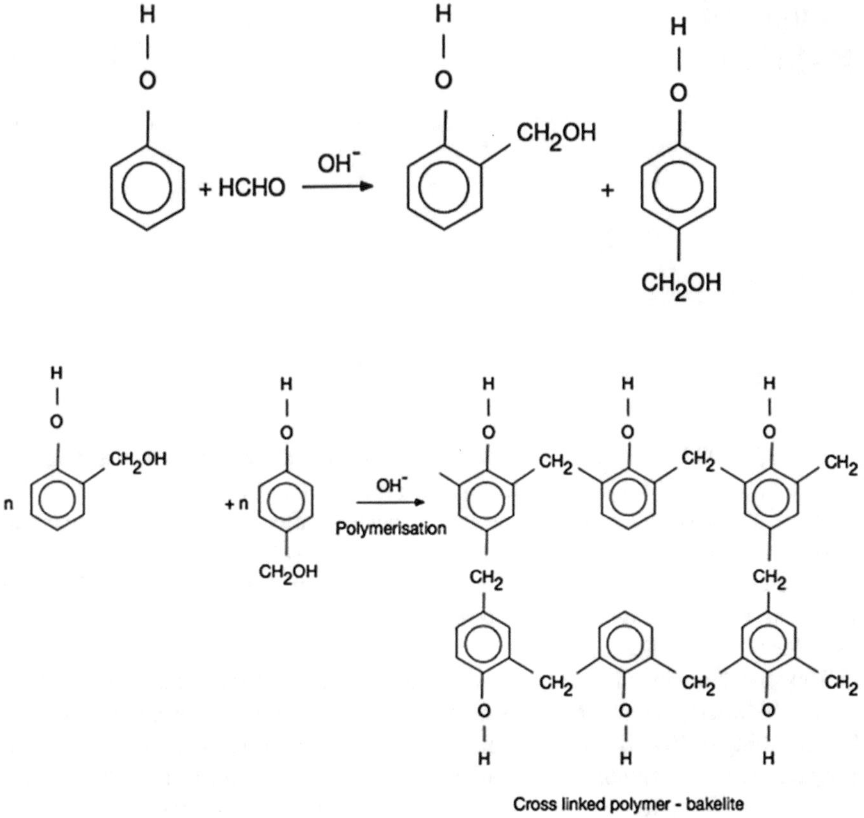

Fig. 4.1 Bakelite formation. (**a**) Initially Phenol and formaldehyde react to form ortho and para hydroxymethyl phenol, (**b**) these are then polymerized using a base (OH$^-$) catalyst at around 120 °C. The formaldehyde to phenol ratio is about 1.5:1. The hexagonal ring has a carbon atom at each corner, which is joined to the two adjacent carbon atoms and a hydrogen atom

Fig. 4.2 A saucepan and a cooking spoon with bakelite handles, which are both common uses for bakelite

Above we noted Bakelite's uses in the past tense. Bakelite is still used today for electrical insulting applications in power generation, electronics and aerospace industries, for some precision components, for various game pieces and for some specialty applications. It is often found as saucepan handles, kitchen knife handles, knobs, and electrical light sockets, see Fig. 4.2. However, in many applications, Bakelite has now been superseded by cheaper, easier to make materials that are less brittle. In fact, some old Bakelite items have become collector's pieces especially vintage costume jewelry items such as necklaces, pins and bangles. Many of these sell for hundreds of dollars.[3] Inevitably, a fake bakelite industry has sprung up in Asia to address this market.[4]

References

1. Weidmann, G., Lewis, P., & Reid, N. (Eds.). (1990). *Structural materials*. Butterworths: The Open University. IBN: 0-408—04658-9.
2. Amato, I. (1998). *Stuff: The materials the world is made of*. New York: Avon Books, Inc.
3. Sass, S. L. (1998). *The substance of civilization*. New York: Arcade Publishing. ISBN: 1-55970-371-7.

---

[3] https://www.thespruce.com/bakelite-jewelry-price-guide-4062356
[4] http://www.angelfire.com/ca3/gale/

Chapter 5
Bronze

Imagine that you are on a battlefield somewhere in the Middle East at the beginning of the Bronze Age. You advance with your copper sword and engage a foe armed with a bronze sword (a copper-tin alloy). He swings, you parry and your sword bends. Hopefully, you have time to retreat and bend your sword straight before you are slain. Whether such a scenario would have happened is debatable. The copper swords would have likely accidentally contained arsenic, which make them harder. Indubitably, bronze swords could keep a sharper edge. In fact, some swords made in China as early as the Warring States period (475–221 B.C.) were engineered to have a high tin content (17–21% tin) along the edge, which makes it harder and better at holding an edge but more brittle, while the center of the sword has a lower tin content (10%) and is softer but more ductile.[1]

At equilibrium, tin atoms are almost completely insoluble in copper. However, the precipitation of other phases is so sluggish that in the low tin content bronzes (<10 weight percent) the tin remains in solution and the bronze is a single phase, see Fig. 5.1. The tin atoms strengthen the bronze because they are a different size to the copper atoms and so cause strain in the lattice when they replace them. Linear defects, called dislocations, when gliding interact with the strain fields around these tin atoms producing so-called *solid solution strengthening*. The arsenic in copper swords has a similar solid solution strengthening effect, but the amount of arsenic in the copper was much less. In contrast, the high tin content edges of the Chinese swords contained so much tin that second-phase particles formed upon casting so that the sword edges were strengthened both by the particles and by solid solution strengthening. Thus, the swords edges were much stronger, but the particles also tend to make them more brittle.

The date of the transition from the Copper Age to the Bronze Age depends on the geographic region. It appears to have started first in the in the west of the Iranian plateau towards the end of the 4th millennium B.C. [1] whereas at the other end of

---

[1] http://www.arscives.com/historysteel/cn.article.htm

© Springer International Publishing AG, part of Springer Nature 2018
I. Baker, *Fifty Materials That Make the World*,
https://doi.org/10.1007/978-3-319-78766-4_5

Fig. 5.1 Schematic showing the close-packed layer of (white) copper atoms in bronze. The tin atoms (black) substitute for the copper atoms

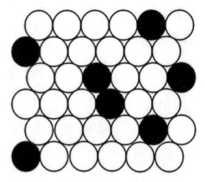

Fig. 5.2 Left, bronze spear from the Eastern Han Dynasty found in Dingxian in 1959; right bronze dagger Found in a tomb in Zhongshan in 1968. Hebei Museum, Changsha, China. (Courtesy of Min Song)

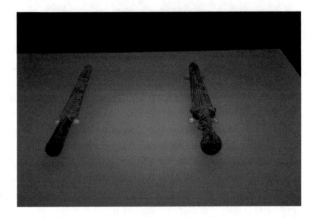

the timescale, both bronze and iron were introduced simultaneously in Japan, but not until around 300 B.C.[2] The Bronze Age was superseded by the Iron Age. This was not because the iron produced was better, in fact bronze is harder than iron (the Vickers hardness of iron is 30–80 versus 60–260 for bronzes) but because iron was more plentiful and easier to process. Cannons were originally made from bronze and even when iron cannons could be made, bronze cannons could be made with thinner walls and were useful for mobile artillery. They were also less likely to burst than iron cannons. The low friction of bronze also means that iron cannonballs are less likely to stick in the barrel.

The end of the Copper Age or Chalcolithic period meant a switch from the use of copper to bronze not only for weapons, but also for bowls, statues and other objects. Bronze is an excellent material for casting and is still extensively used today for statues using the lost wax process. Bronze is strong and ductile and, in addition, exhibits unusual and useful behavior for casting in that the molten bronze expands slightly just before solidifying. This enables the finest details of a mold to be filled

---

[2] Heritage of Japan, Discovering the Historical Context and Culture of the People of Japan, https://heritageofjapan.wordpress.com/yayoi-era-yields-up-rice/the-advent-of-agriculture-and-the-rice-revolution/when-japan-entered-the-iron-age/

Fig. 5.3 Bronze Ding of "Dazhu" from Western Han Dynasty Height: 23.2 cm. Changsha museum, China. (Courtesy of Min Song)

Fig. 5.4 Bronze Kettle Inlaid with Gold and Silver (see right), and with Sealed Character patterns, from the Western Han Dynasty, Height: 44.2 cm; Abdomen Diameter: 28.5 cm Found in the tomb of Zhongshan King in 1968; Hebei museum, Changsha, China. (Courtesy of Min Song)

with the molten bronze. After solidification, like all metals, the bronze shrinks as it cools allowing easy removal from the mold. Ancient Chinese Bronze castings, see Figs. 5.2, 5.3, and 5.4, typical had about 15 wt. % tin [2], which is an excellent composition for casting, indicating that the Chinese clearly experimented with different compositions and could control them. A bronze cauldron (height: 53.8 cm, diameter: 47 cm, weight: 34.7 kg) owned by Duke Mao of the Late Western Zhou Dynasty (1046–771 B.C.) is one of the three treasures of Taiwan's National Palace Museum. These were taken from the Forbidden City in Beijing by Generalissimo Chiang Kai-shek's Nationalist army as part of a large collection of Ancient Chinese artifacts. They were originally evacuated in 1933 to prevent them falling into the hands of the advancing Imperial Japanese Army during the Second Sino-Japanese War. When the Nationalist lost the Chinese Civil War the articles were shipped to Taiwan in 1948 and 1949.

Interestingly, tin and copper are rarely found together. That these two metals were brought together to make bronze indicates the extent of international trade in the Ancient World. For example, tin was transported to the Mediterranean from as far away as Cornwall, England, where there are extensive deposits, from at least

2000 B.C. These Cornish tin mines were worked until the end of the twentieth century when they became unprofitable.

Bronze is 10% denser that steel but does not rust like steel, and is very resistant to seawater, but unfortunately it costs more. Bronze has been used to make bells since at least 1000 B.C. in China [3]. It is also used for cymbals because of both the durability and outstanding acoustic qualities. Bronze is used for springs, bearings and bushings because of its strength, low-friction and corrosion resistance. The earliest bronze coins were made in China in the tenth century B.C. and in India by the sixth century B.C. Bronze was also used to make large monuments such as the fourth century statue of the Greek goddess Athena [3] and the 35 m high Colossus of Rhodes that was erected in 280 B.C. at the entrance to the harbor of the Greek island of the same name [3].

While Ancient bronzes were copper-tin alloys, the modern use of bronze includes copper alloys that have other elements, such as lead, zinc, nickel, bismuth, aluminum and manganese, and may not even contain tin. For example, Commercial Bronze, which is used for architectural applications, is 88 wt. % copper and 12 wt. % zinc, and technically is really a brass. Bronze may not have the significance that it had several thousand years ago, but its use will continue into the future along with the development of new alloys.

References

1. Oudbashi, O., & Davami, P. (2014). Metallography and microstructure interpretation of some archaeological tin bronze vessels from Iran. *Materials Characterization, 97*, 74–82.
2. Amato, I. (Ed.). (1998). *Stuff: The materials the world is made of.* New York: Avon Books, Inc.
3. Emsley, J. (2001). *Nature's building blocks: An A–Z guide to the elements.* Oxford: Oxford University Press. ISBN: 0-19-850340-7.

---

[3] National Geographic.

Chapter 6
Celluloid

Celluloid is a material that was invented to solve a materials shortage in a game – which it didn't really do – but ended up in a enabling a whole new industry.

Billiards and other cue sports started in Northern Europe sometime in the fifteenth century, and by at least the end of the sixteenth century some billiard balls were being made of ivory. Ivory was ideal for this application since it didn't chip, dent or crack, and could be readily made into spherical balls [1, 2]. The ivory could also be dyed to give different colors, although this didn't matter that much since traditional English Billiards only uses three balls, two of which are the white cue balls (one with a black dot) and a red target ball on a six-pocketed table. With the English Industrial Revolution billiards and other cue games proliferated in taverns as the price of manufacturing billiard tables dropped. Later in the U.S.A. 15-ball pool became popular, surpassing four-ball American Billiards in popularity by the 1870s. All this led to great demand for ivory since very few balls, which were solid ivory, could be made from each elephant tusk. Along with the high cost of ivory, the concern that the World would run out of elephants for billiard balls led Michael Phelan in 1863 to offer a prize in the New York Times of $10,000 ($180,000 in 2017 dollars) for the patent rights for an ivory substitute. Phelan (1819–1871) was an Irish-American, who was the first U.S. superstar billiard player, the owner of billiard halls, and the co-owner of Phelan and Collender [3], a New York City company that manufactured patented American billiard tables.

This tempting prize prompted a New York State printer, mechanic and inventor John Wesley Hyatt (1837–1920) to develop celluloid for billiard balls from gum camphor, alcohol and cellulose nitrate, which is produced by adding nitric acid to cellulose from cotton, see Fig. 6.1. The resulting mixture could be molded when heated to 65–150 °C under pressure. In 1868 he and his brother Isaiah started the Albany Billiard Ball Co., and in 1870 he patented the celluloid production process. However, he wasn't awarded the $10,000 prize - unfortunately, celluloid billiard balls didn't play like their ivory counterparts [4]. Worse, his patent was eventually set aside in 1884 after a long legal battle with an English inventor Daniel Spill (1832–1887). Spill had earlier developed a similar product to celluloid called

© Springer International Publishing AG, part of Springer Nature 2018
I. Baker, *Fifty Materials That Make the World*,
https://doi.org/10.1007/978-3-319-78766-4_6

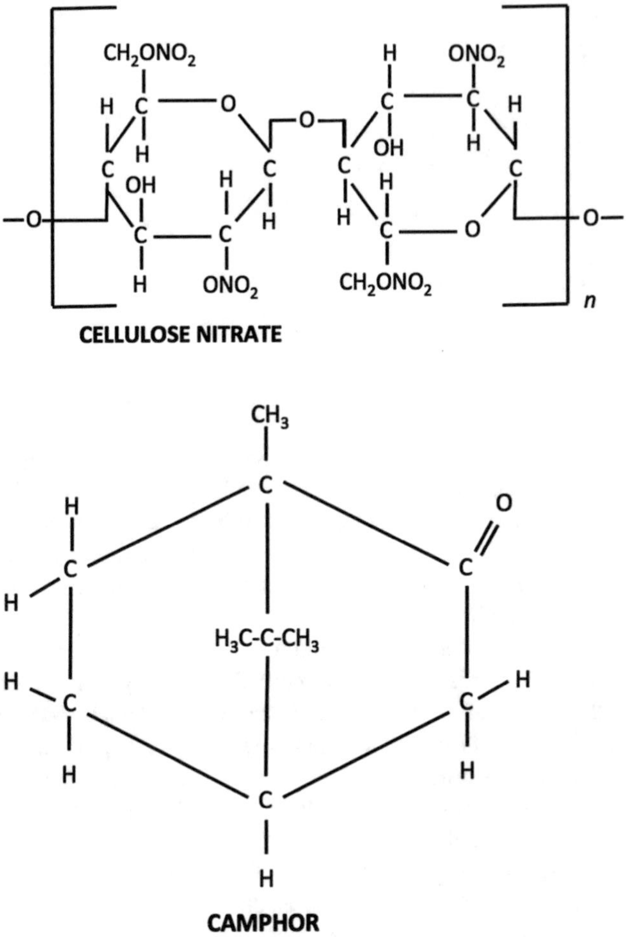

Fig. 6.1 Celluloid is a polymer based on repeating the cellulose nitrate molecule mixed with camphor and alcohol

Xylonite, with the English metallurgist and inventor Alexander Parkes (1813–1890). Previously, in 1855, Parkes had produced the first artificial thermoplastic polymer, Parkesine,[1] from a mixture of cotton fiber, vegetable oils, nitrates, camphor and alcohol.[2] Unfortunately, Parkesine was not a commercial success.[3] While Hyatt's celluloid was not a good solution to the billiard ball challenge, what he had produced was the first commercially-successful artificial plastic, and he and his brother Isaiah started the Celluloid Manufacturing Company in 1871, which made combs, brushes, dentures, cutlery handles and other products where it substituted for ivory.

---

[1] https://www.britannica.com/biography/Alexander-Parkes#ref207559

[2] Plastic Man, Materials World, December, 2017. p. 48–49.

[3] http://www.craftechind.com/the-invention-of-plastic-materials-from-parkesine-to-polyester/

Fig. 6.2 One of the few present uses of celluloid is for guitar picks

Fig. 6.3 Table tennis or ping pong balls are a current use of celluloid. Unlike the apocryphal stories of exploding celluloid billiards balls in the nineteenth century, ping-pong balls have not been reported to explode when struck

It could also be printed to mimic wood or marble. Thus, they gave birth to the first commercially-successful plastics industry. Nowadays, because it is relatively expensive to produce compared to other plastics and highly flammable, celluloid has only a few niche uses such as ping pong balls and musical instruments – where it is used because of its excellent acoustic properties and easy moldability, see Figs. 6.2 and 6.3.

In between celluloid's rapid ascent and its decline in use after other superior, cheaper, and safer plastics were developed, it gave rise to whole new industry. Photographic plates could be produced by spreading cellulose nitrate mixed with silver halide crystals onto glass – light impinging on the crystals turns them black. In 1888, John Carbutt (1832–1905), an English photographer, who had moved to the United States and in 1871 had founded the Keystone Dry Plate Works, coated

Fig. 6.4 A 35 mm 24 shot
cassette with cellulose
triacetate or safety film.
William Kennedy Dickson
developed the 35 mm
format with perforated
edges for celluloid film in
the nineteenth century, but
celluloid is no longer used
because of its flammability

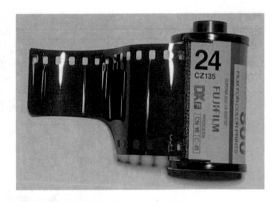

photosensitive gelatin emulsion onto thin celluloid strip.[4] This was used in Thomas
Edison's (1847–1931) Kinetograph,[5] an early camera for capturing motion pictures
invented by Edison employee William Kennedy Dickson [6] (1860–1935), who developed the 35 mm format film with perforated edges, which is still used today (see
Fig. 6.4). In 1889, Kodak employee Henry H. Reichenback (1869–1957) also modified celluloid so that roll film could be produced from it. Meanwhile in 1887,
Hannibal Goodwin (1822–1900), an American episcopal vicar, developed a more
flexible transparent celluloid film, on which to record his biblical teachings, for
which he was eventually awarded a patent in 1898. The patent was acquired by a
Binghampton, NY-based Company Ansco and after several years of litigation they
were awarded \$5 M in 1914 against Kodak Eastman for patent infringement.[7]

Flexible celluloid film and the 35 mm format not only enabled easier photography but also spawned the motion picture industry. However, around 1948 celluloid
was replaced by cellulose triacetate, so-called "safety film" because it did not have
the flammability and occasionally explosive nature of celluloid [5]. Unfortunately,
celluloid undergoes degradation from the environment (moisture, heat, chemical)
and so old celluloid films often deteriorate badly.

References

1. Miodownik, M. (2014). *Stuff Matters*. New York: Houghton Mifflin Harcourt Publishing Co.
   ISBN 978-0-544-23604-2.
2. (1993, November 9). *The bakelizer, National Museum of American History, Smithsonian
   Institution*. Washington, DC: The American Chemical Society, compiled by Vivian Powers.

---

[4] http://www.historiccamera.com/cgi-bin/librarium2/pm.cgi?action=app_display&app=
datasheet&app_id=1782

[5] https://www.thoughtco.com/history-of-the-kinetoscope-1992032

[6] https://www.scienceandsociety.co.uk/results.asp?image=10301013&wwwflag=2&imagepos=1

[7] http://query.nytimes.com/gst/abstract.html?res=990DE0D9173DE633A25754C2A9659C94659
6D6CF&legacy=true

3. Greeley, H. (1872). *The great industries of the United States: Being an historical summary of the origin, growth, and perfection of the chief industrial arts of this country.* Chicago/Cincinnati: J.B. Burr, Hyde & Co.
4. Amato, I. (1998). *Stuff: The materials the world is made of.* New York: Avon Books, Inc. ISBN-10: 0380731533.
5. Ram, A. T. (1990). Archival preservation of photographic film-a perspective. *Polymer Degradation and Stability, 29*, 4. https://doi.org/10.1016/0141-3910(90)90019-4.

Chapter 7
Clay

Clay seems mundane, but it provided essentials for life for our Ancestors in the form of shelter and as food storage containers, see Fig. 7.1. It's use led to the advancement of civilization by providing a medium to record data, which eventually evolved into writing, and for artistic expression. The use of clay by early humans to make containers, cooking pots, counting tokens, musical instruments, bricks and figurines of gods or goddesses [1] occurred throughout the World.

Clay consists of fine particles less than two microns (two millionths of a meter) in size produced by weathering rocks of feldspar (silicate) minerals ($KAlSi_3O_8$ – $NaAlSi_3O_8$ – $CaAl_2Si_2O_8$). The hydrolysis of the feldspars by water during the weathering produces a variety of clay minerals such as kaolinites $Al_2Si_2O_5(OH)_4$, Illites ($K_{0.8}Al_2(Al_{0.8}Si_{3.2})(OH)_2$) and smectites ($Ca_{0.17}(Al,Fe,Mg)_2(Si,Al)_4O_{10}(OH)_2$. nH_2O).[1] Clays consist of alternating layers of tetrahedrally-bonded silicates (SiO_4^{4-}) and octahedrally-bonded aluminates or magnesates, see Fig. 7.2. Clay also contains some organic matter and water. Clays, which can range in color from gray to yellow, are quite pliable due to the water. Once the clay loses its water, either naturally or by firing, the clay becomes hard and brittle.[2]

The earliest known use of clay to make a figurine is of the so-called Venus, a 11 cm high, voluptuous nude female statuette dating from possibly as long ago as 29th millennium B.C. from the Paleolithic, that is hunter-gatherer, Gravettian culture settlement of Dolní Věstonice in the present-day Czech Republic [2]. Along with the Venus figurine were over 2000 burnt clay balls and figures of various animals.

The earliest pottery vessels date to the eighteenth millennium B.C.: fragments of two deep beaker shaped vessels [1] were found in Yuchanyan Cave, Hunan, China[3] along with animal bones, stone tools and ashes indicating it was a camp belonging

---

[1] https://wwwf.imperial.ac.uk/earthscienceandengineering/rocklibrary/viewminrecord.php?mineral=smectite

[2] https://www.sciencelearn.org.nz/resources/1771-what-is-clay

[3] https://www.thoughtco.com/yuchanyan-cave-hunan-province-china-173074

© Springer International Publishing AG, part of Springer Nature 2018
I. Baker, *Fifty Materials That Make the World*,
https://doi.org/10.1007/978-3-319-78766-4_7

Fig. 7.1 One-handled jug
thought to be from the
Eretria region of Greece
from approximately
550 B.C. Otago Museum,
Dunedin, New Zealand

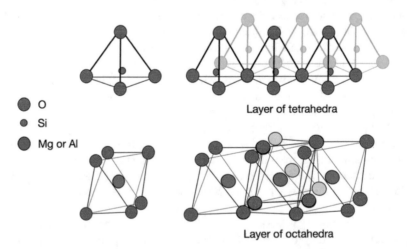

O

Si

Mg or Al

Layer of tetrahedra

Layer of octahedra

Fig. 7.2 Alternating layers of tetrahedrally-bonded silicates (SiO_4^{4-}) and octahedrally-bonded
aluminates or magnesates are the basis of clay

to Late Paleolithic hunters. Pottery dating from as far back as 14,500 B.C. has been found in the Odai Yamamoto I site in Aomori Prefecture, Japan.[4]

For firing in an open hearth, in which the maximum temperature is around 900 °C, early vessels had round bottoms and simple shapes to minimize the thermal stresses that could lead to cracking. Heating clay in this way turns it into pottery by driving off the water and hydroxyl groups, leading to significant shrinkage. Pots were made by shaping by hand or by stacking up coils of clay.

Two great leaps forward occurred around the same time. The potter's wheel was invented around 6000–4000 B.C. in Mesopotamia (modern-day Iraq) and kilns have been found at the Yarim Tepe site also in Mesopotamia dating to roughly 6000 BC [3].

One of the oldest uses of clay was as tokens. The Mesopotamians in around 8000 B.C. used such tokens as part of accounting system for their agricultural products and other goods they produced. Sometime later they collected these in clay containers that were then sealed and the number of tokens in the container was then indicated using a symbol on the outside of the container. The next logical step was to do away with the clay tokens themselves and simply record quantities using the symbols alone. The symbols also evolved from pictures of the objects being recorded to abstract symbols.[5] Exactly when and where writing evolved is still the source of some controversy but it appears to have occurred sometime in the sixth millennium B.C. in the Near East [1].[6,7] Early writing appears to be the simple accounting records of stock or transactions, but clay tablets were involved from the beginning. These tablets were written on with a sharp object, likely a sharpened reed, and then either fired as a permanent record or, if unfired, recycled for future use.[8] Many thousands of clay "documents" have been found in the Near East from the Sumerian civilization in southern Mesopotamia (present-day southern Iraq) including whole clay tablet libraries. Although early writing was in the form of pictograms or hieroglyphs, the cuneiform developed by the Sumerians, and ideograms soon emerged. The rest one can say is history – literally: writing enabled history to be recorded, poems and fiction to be developed and governments to record their business.[9] The first great literary work that we know of Gilgamesh started with five Sumerian poems dating to around 2100 B.C.: several stories in the Bible are similar to the plots in Gilgamesh.[10,11]

A significant improvement in the strength and toughness of pottery was the development of porcelain in China that occurred around 2000 years ago. The

---

[4] http://jomon-japan.jp/en/jomon-sites/odai-yamamoto/

[5] http://www.historyworld.net/wrldhis/PlainTextHistories.asp?historyid=ab33

[6] http://www.ancient.eu/writing/

[7] http://www.ancientscripts.com/sumerian.html

[8] http://www.metmuseum.org/toah/hd/wrtg/hd_wrtg.htm

[9] http://www.skwirk.com/p-c_s-14_u-472_t-1285_c-4932/VIC/8/Government/Ancient-Sumer-Part-A/Ancient-Sumer/History/

[10] https://www.britannica.com/topic/Gilgamesh

[11] https://www.bibleodyssey.org/places/related-articles/gilgamesh-and-the-bible.aspx

Fig. 7.3 Decorative "Bone China" pottery made by Josiah Wedgwood and Sons, a company founded in 1759 in Stoke-on-Trent, U.K. The city is commonly known as The Potteries

Chinese discovered that a clay that incorporated the aluminosilicate clay mineral Kaolin, $Al_2Si_2O_5(OH)_4$, with minerals such as quartz and feldspar when fired at a high temperature produced a material that was not only strong but translucent with a very smooth finish. The effect arises because mullite (either $2Al_2O_3 \cdot SiO_2$ or $3Al_2O_3 \cdot 2SiO_2$) is produced during the high firing temperatures and the result is a vitrified, that is glass, product, which can be glazed or painted. It is so strong that very thin-walled cups and vessels can be produced. European porcelain was not produced until the eighteenth century and the import of these prized Chinese porcelain products into Europe gave use to the word "China" for fine tableware, see Fig. 7.3. China guarded its porcelain manufacturing technology. Eventually, at the beginning of the eighteenth century either the German scientist, physician, and philosopher Ehrenfried Walther von Tschirnhaus (1651–1708) or the German alchemist Johann Friedrich Böttger (1682–1719) developed a European porcelain,[12,13] a development that led to the porcelain industry in the Kingdom of Saxony (now part of modern day Germany) and to Dresden China. The Dresden porcelain collection, founded by the Saxon Prince-Elector Augustus II the Strong (1670–1733) in 1715, is now housed in the magnificent Zwinger Palace in Dresden, and contains over 20,000 pieces of porcelain including fine tableware and figurines, collections of Chinese and Japanese porcelain, and older forms of Dresden pottery called redware.[14] Stoke-on-Trent, known as "The Potteries", has been a center of pottery production in England since the seventeenth century.[15] Probably the most famous producer there is Wedgewood, which was founded in 1759, see Fig. 7.3. Porcelain is also used for dental crowns and dental veneers, see Fig. 7.4.

Bricks have also been used for many thousands of years: the earliest bricks, which date to at least 7500 B.C., were found at Tell Aswad, a 5 hectare (about 12.5

---

[12] http://galileo.rice.edu/Catalog/NewFiles/tscrnhas.html

[13] https://www.britannica.com/art/arcanist#ref167625

[14] http://www.skd.museum/en/museums-institutions/zwinger-with-semperbau/porzellansammlung/history-of-the-collection/

[15] http://www.visitstoke.co.uk/potteries/

Fig. 7.4 Porcelain is used for both dental crowns and dental veneers. (Courtesy of Crystal Wild)

acres) settlement situated on a small river in present-day south-west Syria.[16] The bricks were made by mixing reeds with clay-containing mud and then allowing them to dry. Brick and roof tiles were originally made by drying in the sun. Many such sun-dried bricks were produced in Mesopotamia (a fertile region around the Tigris and Euphrates rivers) and Elam (modern day Iran). The problem with sun-dried bricks is that they degrade quickly. As buildings made of sun-dried bricks degraded, the Ancient Babylonians and Sumerians simply filled in degraded buildings with mud bricks and built on top, producing a multi-layered structure dating back in time called a "tell". Many Mesopotamian cities such as Uruk, Ur and Babylon used sun-dried bricks to build ziggurats, terraced step pyramids, at their centers as a place of worship. The most famous was the 100 m high, 100 m square base ziggurat of Marduk, the patron god of Babylon. Before Ziggurats, temples and shrines in Mesopotamia from 6500–3800 B.C. were built on platforms from sun-dried mud bricks [4]. Kiln-produced bricks were likely first produced in Mesopotamia. These were not only stronger but lighter, and fired bricks are very long lasting being essentially an artificial stone.

The first use of glazing did not occur until the thirteenth century B.C. on bricks for the Elamite Temple at Chogha Zanbil, Iran.[17] The glaze not only made the bricks more attractive but also made them impervious to water. Glazing is now used extensively on pottery.

Modern bricks are no longer made of clay and straw, but are typically 50–60% (by weight) Silica (sand), 20–30% alumina (clay), 2–5% (CaO) Lime, up to 7% iron oxide with a little (less than 1%) magnesium oxide. Surprisingly, it was not until the 1850s that the first brick-making machines were patented: Richard A. Ver Valen (1795–1876) of Haverstraw, New York obtained a patent for an automatic brick

[16] http://www.abovetopsecret.com/forum/thread819706/pg1

[17] http://arthistorysummerize.info/ArtHistory/glazed-bricks/

machine in 1852 that revolutionized brick-making in the U.S.A [18]; Bradley & Craven Ltd. of Wakefield England obtained a British patent for a "Stiff-Plastic Brickmaking Machine" in 1853 and sold the machines throughout the U.K [19]; while Henry Clayton, of Dorset Square, Middlesex, England, obtained a U.S. patent for a Brick and Tile Machine in 1855 that was used at the Atlas Works, London to produce the revolutionary amount of up to 25,000 bricks per day and required only two men and four boys to operate [5].

While the use of clay to produce bricks and pottery continues to grow, clay is finding new uses. Clay animation was first used in films as early as 1908 by the Edison Manufacturing that was started by inventor and entrepreneur Thomas Alva Edison (1847–1931). It use has taken off in both movies and computer games. A more recent development is in the clay/polymer composites, materials that were invented in the 1980s. These are now progressing to clay/polymer nanocomposites that contain clay nanoparticles that have been used to significantly increase the strength and elastic modulus of polymers such as nylon and polypropylene at relatively low cost [6]. The use of calcined clays in concrete is also being explored since less fly ash, which is often incorporated in concrete, is being produced since less coal is being burnt for power [7].

References

1. Sass, S. L. (Ed.). (1998). *The substance of civilization materials and human history from the stone age to the age of silicon*. New York: Arcade Publishing. ISBN-13: 978-1611454017.
2. Amato, I. (1998). *Stuff: The materials the world is made of*. New York: Avon Books Inc. ISBN-10: 0380731533.
3. Streily, A. H. (2000). Early pottery kilns in the Middle East. *Paléorient, 26*, 69–81.
4. Chaline, E. (2012). *Fifty minerals that changed the course of history*. Buffalo: Firefly Books Ltd.
5. Brooman, R. A. and Reed, E. J. (1859, December 9) *The Mechanics' Magazine and Journal of Engineering, Agricultural Machinery, Manufactures, and Shipbuilding, 2*(50).
6. Gao, F. (2004, November). Clay/polymer composites: The story. *Materials Today, 7*(11), 50–55.
7. The rise of calcined clays. *Materials Word*, January 2018, pp. 36–38.

---

[18] https://www.findagrave.com/cgi-bin/fg.cgi?page=gr&GRid=100689862

[19] http://www.gracesguide.co.uk/Bradley_and_Craven

Chapter 8
Concrete

If you ask most people what the most commonly used material is, they might say wood, or steel, or aluminum. The correct answer is actually concrete, which is used in larger quantities than the combined weight of all metals used in a year [1]. Twice as much concrete is used as all other building materials - three tonnes per person are used annually.[1] Another common belief about concrete is that when concrete is being used it has to "dry out". In fact, concrete is a composite material consisting of rock and sand that are "glued" together by cement which itself undergoes a complex set of reactions with water during which it hardens.

Concrete's current predominance as a structural material arises because of its low cost, excellent strength in compression, durability, resistance to weathering and erosion, and its 100-year service life. 85% of the energy cost of a building is running it with only about 15% of the energy used in producing the building materials and its construction. Similarly, 80% of the CO_2 emissions arise from running a building.[1] The use of concrete with its high thermal mass means that buildings need less heating and cooling leading to life-cycle savings of 20% or more compared to some other building materials [2] Concrete is also a safe building material since it does not burn. Light colored concrete in roads and buildings reflects heat more than darker materials minimizing urban heat islands as well.

While concrete seems to be a ubiquitous feature of the modern age featured in buildings, bridges, and also in roads, concrete and cement are not new materials. Their earliest use appears to be by the Nabataeans, a people who inhabited parts of what is current-day Jordan and Syria from around 6500 B.C. They successfully combined lime, produced in kilns, with surface deposits of fine silica sand and water to create waterproof cement that can be seen in some Nabataean structures that survive to this today [2]. The utilization of cement or concrete has also been established for the Ancient Egyptians, Creteans, Greeks, Chinese and peoples in the former Yugoslavia, but the master concrete builders of the Ancient world were the

---

[1] Sustainability benefits of concrete, http://www.wbcsdcement.org/index.php/en/about-cement/benefits-
[2] http://www.concretesask.org/resources/why-is-concrete-better

© Springer International Publishing AG, part of Springer Nature 2018
I. Baker, *Fifty Materials That Make the World*,
https://doi.org/10.1007/978-3-319-78766-4_8

Romans [2]. Previous civilizations had mostly used, mud, clay, natural stone or wood for buildings. In contrast, numerous Roman buildings were constructed using concrete often faced with other materials such as stone, brick or marble. Two of the most famous and still standing buildings in Rome, the Coliseum (82 AD) and the Pantheon (125 AD), contain large amounts of concrete, and the Pantheon still has the record for the largest freestanding dome (43 m in diameter) made of unreinforced concrete. Under the Emperor Claudius, the Romans started to build aqueducts from concrete. That the Romans built things to last is attested by Roman marine concrete. This concrete, which was hydrated using seawater, was used in harbors and can still be found in excellent condition. In fact, so far attempts to hydrate cement or mortar using seawater have not been able to reproduce the excellent marine concrete developed by the Romans [3]. Perhaps, starting with the Romans, we might be said to live in the Concrete Age. Some modern architectural masterpieces are also made of concrete such as the Sydney Opera House, which uses thin concrete shell construction to provide its unique look.

Early Roman concretes contained non-hydraulic lime cement, which sets by carbonation through a reaction with carbon dioxide in the atmosphere. This hardening process happens slowly and produces relatively weak cement.[3] The lime mortar can be produced by heating chalk, seashells and other materials to over 900 °C. The Greeks used such a process to produce mortars that they mixed with rocks from Santorini [4] - an island in the Aegean Sea that was formed volcanically - to produce a high-strength, water-resistant material. From around 200 B.C., the Romans, using a similar approach, started adding volcanic ash from Pozzuoli near Naples in Southern Italy to produce concrete. The pozzolanic ash, which consists of silica, alumina and small amounts of iron oxide, reacts with the slaked lime, $Ca(OH)_2$, and water to produce calcium silicate hydrate and aluminum silicate hydrate that bond strongly with stones and rocks added to produce the aggregate. This cement, which can set under water and does not require the carbonation reaction noted above, is referred to as hydraulic cement. It is much stronger than the lime cements and concretes and can be used underwater. The Romans also demonstrated their technical prowess by producing artificial pozzolans by heating kaolitic clay and volcanic stones to an elevated temperature, an approach called calcining [2]. After the collapse of the Roman Empire when Western Europe entered the Dark Ages, it was also a Dark Age for concrete and cement. Concrete of a similar quality to that produced by the Romans was not produced again for over 1000 years. However, concrete was still used and can be seen in a number of buildings including in concrete technology introduced by the Normans to England that was widely used in castle building [3].

The most commonly used cement today is a hydraulic cement called Ordinary Portland Cement, OPC. Modern cement development was initiated in mid-eighteenth century Britain by John Smeaton (1724–1792), in which, by experimentation, he found that cements containing large portions of clay would harden under water. Using his cement and additives such as pozzolanas and trass, a light-colored volcanic ash that resembles pozzolana, he rebuilt the Eddystone lighthouse,

---

[3] The History of Concrete and the Nabataeans, http://nabataea.net/cement.html

sometimes known as Smeaton's tower, using concrete in 1759. The hydraulic cement used by Smeaton set relatively rapidly, but was not very strong.

In 1791, the Reverend James Parker was granted a patent entitled "Method of Burning bricks, Tiles, Chalk" for what he called Roman Cement - which is actually nothing like cement made by the Romans – that was based on burning natural stones, grinding them up and adding water.[4,5] He built a factory in Northfleet, Kent, England where batches of concrete were made in bottle kilns, which were lined with bricks. Later, Parker sold his patent to Samuel Wyatt, who manufactured cement under the name Parker and Wyatt.

In 1824, Joseph Aspdin (1778–1855), an English bricklayer, patented Portland cement, so called because of its resemblance to a building stone found on the Isle of Portland, Dorset, England. Although this cement differs somewhat from the OPC used today, his son William Aspdin (1815–1864) developed what is essentially the modern OPC and produced it in a number of factories. Earlier cements had been made by heating to relatively low temperatures of about 800 °C. OPC is made by calcining at 1450 °C either finely ground powders or a slurry of calcium silicates (at least two-thirds by weight), calcium aluminates, iron oxide and magnesium oxide and few other compounds. The sources of these compounds include limestone (the most abundant mineral), chalk, shells, silica sand, blast furnace slag, fly ash (from coal-fired power plants) and clay. The key reaction is that of Ca_2SiO_4 (belite) with CaO to form Ca_3SiO_5 (alite). The result is clinker, which consists of 5–25 mm chunks containing alite (~50%), belite (~25%), a ferrite phase $4CaO \cdot Al_2O_3Fe_2O_3$ (~10%), and tricalcium aluminate $3CaO \cdot Al_2O_3$ (~10%). This is mixed with gypsum ($CaSO_4$) (5%) and sometimes limestone ($CaCO_3$) and ground to a fine powder with sizes in the range 2–80 microns [5] (a micron is a millionth of a meter) with 95% of the powder particles less than 45 microns in diameter to produce cement. Cement is potentially dangerous. It is highly alkaline (pH ~ 11) and can cause skin burns when wet, and the reaction between cement and water is highly exothermic.

While initially cement was made in batches, the invention of the rotary kiln by the Englishman Frederick Ransome (1818–1893) in 1873 turned the making of cement into a continuous process in which raw materials are fed into one end of a refractory brick-lined rotating cylinder that is slightly inclined to the horizontal and the clinker emerges from the other end.

Concrete is an aggregate composite in which large and small particles reinforce a matrix derived from the reaction of cement with water. The large particles are coarse gravel or rocks typically ranging from a few millimeters in size to a few tens of millimeters, and fine particles of sand, with a range of sizes whose average grain size is a less than a millimeter. The two different size ranges of the aggregate are important since the fine particles (sand) fill in the spaces between the large particles and, thus, very efficiently fill the space [1]. Cement is made up of calcium silicate fibrils that mesh together and bond to the rock and stone, and, thus, acts as the glue for the aggregate, which also includes entrained air. Increasing the amount of

[4] http://www.cementkilns.co.uk/roman.html

[5] http://www.cementkilns.co.uk/cemkilndoc006.html

Fig. 8.1 Rebar in
concrete. The concrete was
chipped away and new
rebar put in place followed
by concrete to repair a
bridge support in Lebanon,
New Hampshire, U.S.A

entrained air, through the addition of various admixtures, can improve the workability of concrete during the forming process and also improves its durability in cold climates that have freeze-thaw cycles.

The strength of concrete depends on a number of factors including the amount of porosity (more porosity decreases the strength), the water/cement ratio (higher water cement ratios decrease the strength), the composition of the cement, and the strength of the rock aggregates in the concrete. Very high strength concretes can reach compressive strengths of 130 MPa, but, like many ceramics, the strength in tension is typically less than a tenth of the compressive strength. Thus, one cannot make load-bearing beams out of concrete because one half of such a beam will be subject to tensile forces. The solution is to reinforce the concrete. The first to reinforce concrete may again have been the Romans, who added horsehairs, which increase the durability and prevent crack propagation. Fiber reinforcement is still used in some concrete, but these days the fibers are made of steel, glass or polymer.

Iron rods were first used to reinforce concrete in house construction by a Frenchman Edmond Coignet (1850–1915) and a Briton William B. Wilkinson (1819–1902) around 1854. Steel rods in concrete produce a composite such that most of the load is carried by the steel. Two features make steel-reinforced concrete possible. First, concrete reacts with steel rebar and bonds to it, see Fig. 8.1. If it didn't, the reinforced concrete would simply have steel rods sliding around inside hollow tubes within the concrete. Second, perhaps surprisingly, steel and concrete have almost the same coefficient of thermal expansion ($8\text{--}10 \times 10^{-6}/°C$), which means that as the reinforced concrete heats up and cools down, the steel and concrete expand and contract the same amount. If they didn't, the bonds between the steel and concrete would be subject to thermally-induced stresses, which would lead to breaking of the bonds between the concrete and the steel and to cracking of the concrete. It is also helpful that concrete is very alkaline so that concrete produces little corrosion of the steel reinforcing rods.

Concrete is somewhat self-healing: cracks in concrete admit rainwater, which with dissolved carbon dioxide from the atmosphere, can lead to carbonate formation and help to close cracks. New ways of producing self-healing concrete are also being explored. Researchers at Delft University of Technology in the Netherlands purposely incorporated bacteria into concrete along with calcium lactate for the bacteria's lunch. The bacteria remain dormant in the concrete until a crack forms and allows in water. The water and calcium lactate enable the bacteria to germinate, multiply and produce calcite, which heals the cracks.[6]

The production of cement accounts for 2–3% of the world's energy consumption each year and generates over 5% of anthropogenic carbon dioxide generation, with about 60% of this carbon dioxide arising from the calcination process itself with the rest from the use of fossil fuels for heating [6]. Thus, 822 kg of carbon dioxide are produced for every tonne of cement, with half of that coming from the heating. Even so, cement production is quite energy efficient typically requiring 2.2 GJ/ tonne, while the theoretical energy requirement is only slightly larger at 3.06 GJ/tonne [7]. The carbon dioxide emission can, of course, be reduced by using non-fossil fuels, and by modifying the composition of the cement and using some limestone in concrete. Another possibility is to turn the carbon dioxide generated into something useful. For example, Blue Planet Ltd., a company based in California, produces calcium carbonate aggregates made from sequestered carbon dioxide using biomineralization, that is, using living organisms.[7]

Although concrete and cement have been used in one form or another for centuries, innovations continue even beyond self-healing. Self-cleaning concrete is one example. It has a photocatalytic additive such as titanium dioxide or zinc oxide that when exposed to the ultraviolet radiation in sunlight breaks down organic particulates and airborne pollutants such as nitrous and sulfuric oxides into carbon dioxide, oxygen, water, sulfates, nitrates and organic compounds that are then washed off the hydrophilic surface by rain [8]. The self-cleaning concrete may also clean the surrounding air, but of course, the pollutants that are washed off end up in the ground water. Another innovation is pervious concrete, that is sponge-like due to a network of voids in the structure, which allows water to penetrate, thus allowing stormwater management when used in car parks and other surfaces.[8]

---

[6] http://www.cnn.com/2015/05/14/tech/bioconcrete-delft-jonkers/

[7] http://www.blueplanet-ltd.com/

[8] What Makes Concrete A Sustainable Building Material? By Anne Balogh.

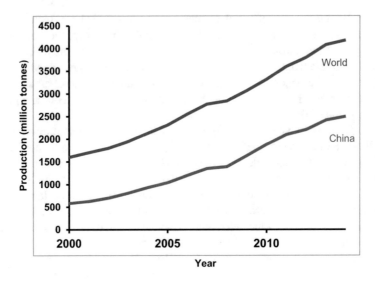

Fig. 8.2 Hydraulic cement production by year[8]

Cement production has been growing rapidly in recent years from 1.6 billion tonnes in 2000 to around 4.2 billion tonnes in 2014 with China's production growing by over 400% from 583 million tonnes in 2000 to 2.5 billion tonnes in 2014 and currently accounting for around 60% global production.[9] (Fig. 8.2). Global production is such that the U.S. State of Rhode Island and Providence Plantations could be paved over with about half a meter of concrete each year. In contrast, the U.S. production at 83 million tonnes is slightly less than it was in 2000 (90 million tonnes). In fact, from 2014–2016, China produced over 50% more cement (7.2 million tons) than the U.S. produced in the whole twentieth century.[10] And amazingly, more than half of the over four billion tons of concrete produced each year is used in China currently.[11] While most developed countries are showing declining production, quite rapid growth is occurring in developing countries such that cement production can be taken as a measure of their growth.

---

[9] U.S. Geological Survey, Mineral Commodity Summaries – Cement .

[10] BBC Podcast: More Or Less: Behind the Stats, "The Concrete Fact's about Trump's Wall and China", March 16th, 2017.

[11] U.S. Geological Survey, Mineral Commodity Summaries, January 2015; http://minerals.usgs.gov/minerals/pubs/commodity/cement/

Side Box

The reaction of cement with water is complex and not fully understood, which is perhaps not surprising giving the large number of ingredients. The heat output from the reaction is also complex, see Fig. 8.A. While the initial reaction with water occurs with alite and the tricalcium aluminate, most of the heat comes from the reactions of water with the alite and belite:

$$\text{Alite reaction}: 2\,Ca_3SiO_5 + 7\,H_2O \rightarrow 3\,CaO.2\,SiO_2.4\,H_2O + 3\,Ca(OH)_2$$

$$\text{Belite reaction}: 2\,Ca_2SiO_4 + 5\,H_2O \rightarrow 3\,CaO.2SiO_2.4\,H_2O + Ca(OH)_2$$

The first reaction releases 174 kJ and the second 58.6 kJ, but both produce the same products of calcium silicate hydrate and calcium hydroxide. The rate at which this reaction occurs depends on the fineness of the cement particles. Finer cement has a larger surface area and reacts faster. In fact, different grades of cement with different particle sizes can reach their maximum strength at times ranging from a day to three months [9].

The reaction of cement with water produces a complex microstructure. High-resolution transmission electron microscope images of hydrated cement show a fibril or crumpled foil morphology.[12] After hydration in the bulk, cement typically consists of both fully hydrated cements grains and residual unhydrated grains surrounded by a hydration shell sitting in a gel of hydration products consisting of calcium silicate hydrate, and calcium hydroxide along with small amount of ettringite or hexacalcium aluminate trisulfate hydrate ($(CaO)_3(Al_2O_3)(CaSO_4)_3\cdot32H_2O$) and monosulfate ($3CaO\cdot Al_2O_3.$ $CaSO_4.12H_2O$) [4]. The ettringite is present as long crystals. The structure of concrete is essentially similar but, of course, small sand grains and large stones are present, and there are pores containing air, which can range in size from 20 microns to 1 mm and constitute 15% of the material [4].

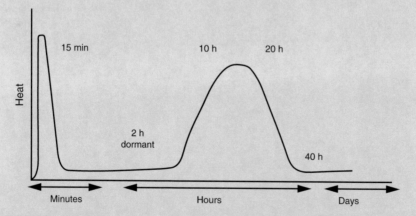

Fig. 8.A Heat evolution as a function of time upon adding water to cement

[12] Concrete Microscopy Library.

References

1. Shackelford, J. F. (2004). *Introduction to materials science and engineering.* ISBN 0–13–142486-6. Upper Sadle River, NJ: Prentice Hall.
2. Gromicko, N., & Shepard, K. *The history of concrete.* http://ezekiel31army.blogspot.com/2016/01/the-history-of-concrete.html
3. Jackson, M. D., Mulcahy, S. R., Chen, H., Li, Y., Li, Q., Cappelletti, P., & Wenk, H.-R. (2017). Phillipsite and Al-tobermorite mineral cements produced through low-temperature water-rock reactions in roman marine concrete. *Maerican Minerologist, 102,* 1435–1450.
4. Sass, S. (1998). *The substance of civilization.* New York: Arcade Publishing ISBN-10: 1611454018.
5. Diamond, S. (2004). The microstructure of cement paste and concrete – A visual primer. *Cement and Concrete Composites, 26,* 919–933.
6. Kurtis, K. M. (2015). Innovations in cement-based materials: Addressing sustainability in structural and infrastructure applications. *MRS Bulletin, 40,* 1102–1107.
7. The rise of calcined clays. *Materials World,* January 2018, pp. 36–38.
8. Zailan, S. N., Mahmed, N., Al Bakri Abdullah, M. M., & Sandu, A. V. (2016). Self-cleaning geopolymer concrete - _A review, International Conference on Innovative Research 2016 - ICIR Euroinvent 2016 IOP Publishing IOP Conference Series. *Materials Science and Engineering, 133,* 012026. https://doi.org/10.1088/1757-899X/133/1/012026.
9. Flinn, R. A., & Trojan, P. K. (1990). *Engineering materials and their applications.* ISBN: 0–395–43305-3. Boston, MA: Houghton Mifflin.

Chapter 9
Copper

Copper is a commonly-encountered metal and most people would easily recognize a piece. So the perception may be that it is common metal, but it is not. It comprises only 0.0068% of the Earth's crust. By comparison, iron and aluminum are 6.3% and 8.1% of the Earth's crust, respectively.[1] The name copper comes from the Latin name *aes cyprium* or the "Cypriot metal", which later became *cuprum* – Cyprus being a major source of copper in Roman times. Elemental copper, which is a reddish-orange color face-centered cubic metal (like aluminum), is one of only four colored metals – the others are the bluish element osmium, and the yellow-colored gold and cesium.

Since elemental copper is one of three metals along with gold and iron[2] that can be commonly found naturally it was also one of the first three metals to be used by man. The first undisputed site in which naturally-occurring copper was made into various artifacts is at Çayönü in Southeastern Turkey, which dates to the late eighth millennium [1]. Copper is very ductile and malleable, and pieces can be beaten into shape, which makes it harder (a phenomenon called work-hardening), and, thus, intermediate heat treatments are used to soften the metal so that it can continue to be worked. Although elemental copper occurs naturally, there is not a lot lying around. However, a 381 tonne slab of copper was found in the Keweenaw Peninsula, the northernmost part of the Upper Peninsula in Michigan, U.S.A. in 1857 [2]. Since copper's melting point is 1084 °C it can be melted in a charcoal furnace and cast much more easily than iron, which melts at 1538 °C. The discovery that heating the green monoclinic-crystal structured mineral malachite (copper carbonate hydroxide, $Cu_2CO_3(OH)_2$) or a copper sulfide or copper oxide ore with charcoal produced elemental copper must have been a revelation – heating the sulfide ores was likely not a pleasant experience because of the sulfur smell. This processing approach is possible since copper is not as strongly bonded in its ores as a metal

---

[1] http://periodictable.com/Properties/A/CrustAbundance.an.html

[2] https://www.webelements.com/copper/

© Springer International Publishing AG, part of Springer Nature 2018
I. Baker, *Fifty Materials That Make the World*,
https://doi.org/10.1007/978-3-319-78766-4_9

Fig. 9.1 A typical
plumbing use of copper

such as aluminum. The smelting of copper probably first occurred around 5000 B.C. in southern Anatolia [1].

There were a wide variety of early uses of copper in the Ancient World including tubing to carry water, ornaments, mirrors, knives, drinking vessels and bowls. Using the lost wax process, in which a mold is initially made out of wax, a process still used today, intricate copper objects could be produced such as crowns that have been dated to 3700 B.C. [3]. Copper was also used to make tools such as axes and chisels, and weapons such as arrowheads and swords. However, copper is not very strong and, thus, pure copper swords bend easily and chisels will quickly lose their edge [4]. Some copper ores contain arsenic and the arsenic accidentally incorporated into the copper would have strengthened the copper substantially, but its vapors would not have been good for the copper worker. Copper swords would have rapidly gone out of use once bronze - which is made by adding about 10% tin to copper - swords appeared. The period after the Neolithic or New Stone Age is referred to as the Chalcolithic or Copper Age and it ended with the introduction of bronze and the Bronze Age.

The main modern uses of copper as an elemental metal are based on its four outstanding properties: high thermal conductivity (401 W/m.°C); high electrical conductivity (5.9×10^7 Siemens/m)[3]; good corrosion resistance, and its antibacterial and antifouling properties. Copper's thermal conductivity and electrical conductivity are second highest of any metal, and close to those of silver, the metal with the highest thermal and electrical conductivities. It is used rather than silver because it is much cheaper. Modern applications of copper generally use the metal in high purity form (99.9%) for electrical wiring (60% of usage), roofing, architectural uses and plumbing (20%), electrical machinery (15%) with the other 5% of copper used for incorporation in alloys, see Figs. 9.1 and 9.2. Copper is used in whisky stills, not only because of it high thermal conductivity but also because it reacts with sulfur compounds removing them and making better tasting whisky. Copper has also long been used on ships to stop bacteria, barnacles and mussels growing on them. However, currently Muntz metal, which is roughly 60% copper and 40% zinc with

---

[3] http://www.tibtech.com/conductivity.php

Fig. 9.2 A lens for an electron microscope showing numerous copper wires used to produce a magnetic field when a large electrical current is passed through them

Fig. 9.3 The U.S. penny is a zinc disc coated with copper

a little iron is mainly used for this application. Copper's antibacterial activity has been extended into new applications in recent years.

Copper is the basis of a number of structural alloys when combined with various other metals: copper + zinc = brass, which is used in a variety of applications including locks, gears and bearings (where its low friction is utilized), doorknobs (for its color), valves and bullet casings, in brass musical instruments, and for some plumbing and electrical applications; 88% copper + tin = gunmetal, sometimes called "red brass", was used to make guns, but this use has been taken by various steels; 66–90 atomic percent copper + nickel = cupronickel alloys, which are used, for example, in the U.S. "nickel" coin; approximately 30% copper + nickel = Monel alloys, which is used for its oxidation and corrosion resistance and good high and sub-zero mechanical properties; and copper +0.5–3 at. % beryllium = Spring Copper, which is both ductile and the highest strength copper alloy at 1400 MPa, which is used in a variety of niche applications such as for percussion instruments, and, as the name suggests, for springs. Copper is also added to gold and silver to modify their colors and increase their hardnesses.

Copper and its alloys produce protective films, which make them resistant to corrosion in water. Thus, copper has been used for making coins for millennia and copper is still used for coins themselves or for coating coins. For example, the U.S. penny has been a zinc disk coated with copper since 1982, see Fig. 9.3.

Copper is also used for decorative purposes – the Statue of Liberty, designed by the French sculpture Frédéric Barthodi (1834–1904) and built by the French engineer and architect Gustave Eiffel (1832–1923), a gift from the French people to American people dedicated in 1886 is the world's largest copper statue and shows

Fig. 9.4 Copper
production as a function of
time

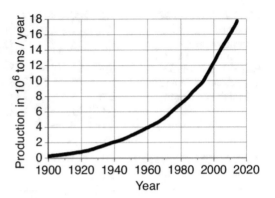

the characteristic green patina.[4] Patinas on copper can be carbonates, sulfides, or sulfates depending on the environment.

One third of copper ores are produced in Chile, whose production accounts for one fifth of its economy. Over 25% of copper is smelted in China. China also utilizes half of the world's scrap copper.[5] Although copper has been used for over 10,000 years, almost half the copper used was consumed in the last 25 years, and as shown in Fig. 9.4, the production of copper continues to grow rapidly. So we might expect the use of copper to continue to grow in the future.

As with many pure metals, pure copper is not very strong with yield strength of only about 70 MPa. However, it has recently been demonstrated that if the grains or individual crystals of which the copper is composed are reduced to the nanosized range then the strength can be dramatically increased. For example, for pure copper with grains that are only 23 nm (23 billionths of a meter) a yield strength of almost 800 MPa can be obtained [5]. Thus, future applications which utilize either the high thermal and electrical conductivities or anti-fouling properties combined with the high strength can be envisaged.

In February 2018, the price of copper at around $7000 per tonne was similar to what it was 10 years earlier, but significantly more expensive than aluminum at $2700/tonne.

References

1. Muhly, J. D. (1988). The beginnings of metallurgy in the old world in the beginnings of the use of metals and alloys. In R. Maddin (Ed.), *The beginning of the use of metals and alloys*. Boston: MIT Press. ISBN: 13 9780262132329.
2. Emsley, J. (2003). *Nature's building blocks: An A-Z guide to the elements*. Oxford: Oxford University Press. ISBN-13: 978-0198503408.

---

[4] Landmarks Preservation Commission, September 14, 1976, number 1, LP-0931.

[5] The Economist, August 5th, 2017.

3. Maddin, R. (Ed.). (1988). *The beginning of the use of metals and alloys*. Boston: MIT Press. ISBN: 13 9780262132329.
4. Chaline, E. (2012). *Fifty minerals that changed the course of history, copper*. Buffalo: Firefly Books Ltd..
5. Guduru, R. K., Murty, K. L., Youssef, K. M., Scattergood, R. O., & Koch, C. C. (2007). Mechanical behavior of nanocrystalline copper. *Materials Science and Engineering A, 463*, 14–21.

Chapter 10
Cotton

The innocuous-looking, even delicate cotton boll has had a huge impact on history. Whether the Glorious Revolution of 1688 provided the legal and cultural genesis of the Industrial Revolution in Great Britain is still debated, but technologically the steam engines of Thomas Savery (1650–1715) and Thomas Newcomen (1650–1715), patented in 1698 and developed about 1712, respectively, and later improved in a new design of 1781 by James Watt (1736–1819) were fundamental to the industrial revolution as machines initially to pump water from mines and later to power factories and mills. Improvements in metallurgy and mining were also very important, as was the production of wool and linen, but the mechanization of cotton production was the greatest showcase of the First Industrial Revolution providing gains of factors of 40–50 in output per person. Nearly all of the inventions associated with these early improvements in cotton production were from British inventors such as: the Flying Shuttle by John Kay (1704–1779) in 1733[1]; the roller spinning frame and the flyer-and-bobbin system by Lewis Paul (unknown-1759)[2] and John Wyatt (1700–1766)[3] patented in 1738; the carding machine invented separately by Lewis Paul and Daniel Bourn[4] in 1748; the Drop Box by John Wyatt's son Robert Kay (1728–1802) around 1760[5,6]; the Spinning Jenny by James Hargreaves (1720–1778) in 1764[7]; the spinning frame, later renamed the water frame, by Richard Arkwright (1732–1792) in 1768[8]; and the Spinning Mule by Samuel Crompton (1753–1827) in 1779, see Fig. 10.1. However, a key invention was the Cotton Gin by the American

---

[1] https://www.britannica.com/technology/flying-shuttle

[2] http://spartacus-educational.com/TEXpaul.htm

[3] https://www.britannica.com/biography/John-Wyatt

[4] https://www.revolvy.com/topic/Daniel%20Bourn&item_type=topic

[5] http://www.gracesguide.co.uk/Robert_Kay

[6] https://www.britannica.com/biography/Samuel-Crompton

[7] https://www.britannica.com/biography/James-Hargreaves

[8] http://www.bbc.co.uk/ahistoryoftheworld/objects/RyHIgvgsSeCYGZRl4Ep5RQ

© Springer International Publishing AG, part of Springer Nature 2018
I. Baker, *Fifty Materials That Make the World*,
https://doi.org/10.1007/978-3-319-78766-4_10

Fig. 10.1 A spinning wheel for turning cotton into yarn that would have been used at home rather than in a cotton mill. Woodstock History Center, Woodstock, Vermont, U.S.A

Eli Whitney (1765–1825) [9]that was patented in 1794. Arkwright's water-powered cotton mill in Cromford, Derbyshire, England is credited as being the first modern industrial factory.[10] The success of the British cotton industry is reflected in the fact that cotton products constituted over 40% of British exports in 1784–1786 [1].

Eli Whitney's cotton gin, a machine that separates cotton fiber from cotton seeds, thus replacing much manual labor, made cotton the biggest cash crop in the U.S.A. from the end of the eighteenth century. Cultivation of cotton required large amounts of manual labor, which was supplied mostly by slaves. Cotton, thus, was one part of the goods that comprised part of a slave trade triangle between Africa, the U.S.A. and the U.K., in which textiles, rum, guns and other manufactured goods were shipped from England to Africa, slaves were shipped from Africa to the U.S.A., and sugar, tobacco and cotton were shipped from the U.S.A. to England. In 1807, driven by the efforts of the evangelical Christian William Wilberforce that were commemorated in the movie "Amazing Grace", the British Parliament passed an act entitled "An Act for the Abolition of the Slave Trade", which ended the slave trade in the British Empire.[11] In the same month, the U.S.A. enacted a similar law "Act Prohibiting Importation of Slaves" that ended the slave trade to the U.S.A., but it did not curtail slavery within the U.S.A. [2]. From then on, the Royal Navy's West Africa squadron intercepted any slave-carrying ships, which it considered the same as pirates.

---

[9] https://www.biography.com/people/eli-whitney-9530201

[10] http://www.bbc.co.uk/history/historic_figures/arkwright_richard.shtml

[11] http://www.nationalarchives.gov.uk/pathways/blackhistory/rights/abolition.htm

The American South was also involved in the next part of the cotton saga. At the beginning of the American Civil War, cotton from the Confederacy provided 77% of the United Kingdom's 400,000 tons of imported raw cotton for their mills with smaller amounts sent to European countries.[12] In an effort to end the neutrality of the U.K. and France and force them to either recognize the Confederacy or enter the civil war on the Confederacy's side, the Confederacy self-embargoed the export of cotton and even burned 2.5 million bales of cotton. This "King Cotton Diplomacy" did not work since large reserves of cotton had been built up, particularly in the U.K. Nevertheless, despite a Union blockade of Southern ports that eventually destroyed the Confederate economy, some blockade-runners exchanged superior British weapons for cotton.[13] The lack of American cotton had a huge negative impact on the British economy, where at one time one quarter of the population depended directly or indirectly on cotton. Britain was now vehemently opposed to slavery and, thus, both Britain and France invested heavily in Egyptian cotton production.[14] This had unintended consequences. After the American civil war ended, Egyptian cotton was abandoned in favor of American cotton. This devastated the Egyptian economy and led to the declaration of bankruptcy by Egypt in 1876. This resulted in an invasion by Anglo-French-Indian forces, and in 1882 Egypt became a British protectorate.

Cotton has been grown from at least 5000 B.C. in Mexico. In the Indus valley, production of cotton textiles has occurred since 6000 B.C., but British restrictions on cotton imports in 1721 started to transform India from a source of textiles to simply a raw cotton producer that imported British cotton products. Eventually, in the 1920s, Mohandas Gandhi (1869–1948), who believed that the goal of Indian Independence was closely tied to cotton, launched a massive boycott of foreign, particularly British, products, and encouraged Indians to utilize homespun cotton products. This eventually led to Indian independence in 1948. Thus, cotton can be associated with both the rise and the decline of the British Empire.

Cotton has a number of excellent properties: it has good strength, excellent absorbency, is machine washable, dry cleanable, retains colors well and drapes well. Thus, cotton is used to make a range of products such as shirts and T-shirts, jeans and trousers, towels and handkerchiefs, and sheets. It is also blended with other fibers such as rayon, polyester and linen. Originally, paper was largely made from cotton, and it is still used in high-end art paper and some banknotes. Cotton is not only important as a material in its own right, but has been an important source of cellulose, see Fig. 10.2, which is the basis of other materials. For example, cotton was the source of the cellulose utilized to make nitrocellulose that was used to produce celluloid, which was originally used to make photographic film.

The cellulose that makes up 88–96.5% of cotton fiber grows in a complex arrangement, see Fig. 10.3. In the middle of the fiber is the lumen, which is a hollow channel that runs the length of the fiber. When the fiber is growing the channel is

---

[12] http://study.com/academy/lesson/king-cotton-cotton-diplomacy-the-civil-war.html

[13] http://mshistorynow.mdah.state.ms.us/articles/291/cotton-and-the-civil-war

[14] http://www.britishempire.co.uk/maproom/egypt.htm

Fig. 10.2 The cellulose molecule, which is the basis of the cotton fiber

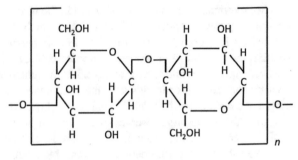

CELLULOSE

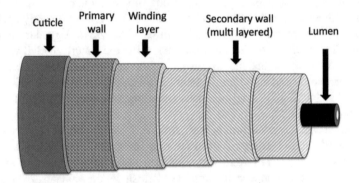

Fig. 10.3 The structure of the cotton fiber

wide and contains proteins, coloring and minerals on the walls, whereas in mature fibers the lumen virtually disappears. Around the lumen are several layers of the secondary wall of almost pure cellulose fibrils at 20–45 degrees to the fiber axis alternating in direction, see Fig. 10.3, that consist of aligned 100 nm (a nanometer is one billionth of a meter) macrofibrils that are themselves composed of 20 nm wide ribbon-like lamellae consisting of 5 nm microfibrils [3]. This arrangement of the secondary walls, which are almost completely made up of cellulose and comprise 90% of the fiber, provides great mechanical strength. The primary wall that encloses the secondary wall consists of 23% cellulose and 22% protein fibrils at around 70 degrees to the fiber axis[15] in an amorphous polysaccharides or sugar matrix of pectin and xyloglucan [4]. Because of this alignment of the cellulose molecules, cotton is about 73% crystalline.[16] The criss-cross arrangement of the cellulose provides resistance to swelling. Surrounding the whole fiber is the cuticle that consists of fats and waxes, sugars (pectins), and minerals that protect the underlying fibers from mechanical and chemical damage. Cotton is mostly used for clothing, but also for a variety of other products such as rope, Fig. 10.4.

---

[15] https://sites.google.com/site/textileandfashiontechnology/letter/natural-fibers

[16] https://www.barnhardtcotton.net/technology/cotton-properties/

Fig. 10.4 Cotton cloth and cotton rope

China and India, which produce similar amounts of cotton, are currently the top two producers at about 6.5 million tonnes,[17] with the U.S.A running third at 3.5 million tonnes, with the U.S.A. being the top cotton exporter. Total cotton production is about 25 million tons, which is usually measured in bales of 500 lbs. (227 kg). 45.6% of textiles are produced from natural fibers with cotton accounting for 39–40% of total textile production by volume.[18] Cotton production occupies 2.5% of the arable land in the World. Unfortunately, the cotton plant is very thirsty and uses large amounts of water. 73% of the global cotton production involves irrigation, which in some places has caused a large fall in water tables [5]. One egregious example is the fate of the Central Asian Aral Sea that once covered an area of 66,000 km^2 and was the fourth largest inland sea. In the 1960s, the Soviet government diverted the rivers Amu Darya and Syr Darya from the Aral Sea in order to irrigate crops, particularly cotton, referred to as white gold, which were grown in the desert. The consequence was that that in less than 25 years the Aral Sea shrank to 10% of its former size. Because of the water needs, it is not clear that cotton production is truly sustainable at present levels.

References

1. Schoen, B. (2009). *The fragile fabric of union: Cotton, Federal Politics, and the global origins of the civil war* (pp. 26–31). Baltimore: Johns Hopkins University Press.
2. U.S Congress: Act to Prohibit the Importation of Slaves, Passed on March 2, 1807.
3. Peterlin, A., & Ingram, P. (1970). Morphology of secondary wall fibrils in cotton. *Textile Research Journal, 40*, 345–354.
4. Meinert, M. C., & Delmer, D. P. (1977). Changes in biochemical composition of the cell wall of the cotton fiber during development. *Plant Physiology, 59*, 1088–1097.
5. Chapagain, A. K., Hoekstra, A. Y., Savenije, H. H. G., & Gautam, R. *The water footprint of cotton consumption, UNESCO-IHE institute for water education, value of water research report series* (No. 18, 2005).

---

[17] http://www.worldatlas.com/articles/top-cotton-producing-countries-in-the-world.html

[18] http://www.grandviewresearch.com/industry-analysis/textile-market

Chapter 11
Diamond

"Diamonds are Forever" according to the 1956 Ian Fleming novel, which was made into a film in 1971, about the British secret agent James Bond. Diamonds may last an awfully long time but they are not forever. Diamond is a metastable (long-lived, but not equilibrium) crystalline form of carbon that has a crystal structure referred to as diamond cubic, which is also adopted by both silicon and germanium, see Fig. 11.1. The stable or equilibrium form of carbon is graphite, the stuff in the middle of your "lead" pencil. However, not to worry, even though diamonds are only metastable, if kept at room temperature it will likely take billions of years before they turn into graphite. On the other hand, they will burn if heated to 850–1000 °C generating the far less valuable greenhouse-gas carbon dioxide. Carbon dioxide production was the key to unraveling both the nature of diamonds and a law of science. Antoine-Laurent de Lavoisier (1743–1794) showed that when either charcoal or diamond were burnt in a closed glass jar they disappeared and the same gas, now called carbon dioxide, was produced. This enabled him to realize that charcoal and diamond were made of the same material, which, he called carbon. That the weight of the glass jar did not change in this experiment also provided evidence for the Law of Mass Conservation that he formulated.[1]

Each natural diamond is a single crystal. The largest diamond ever discovered, weighing in at 3106.75 carats or 0.621 kg (diamonds are traditionally weighed in carats, each of which is 0.2 g), was the Cullinan Diamond found in 1905 in South Africa and named after the chairman of the mine where it was discovered. It was cut into 105 diamonds. The largest of these, named the Great Star of Africa, which weighs 530.4 carats or 0.106 kg, was presented to the British King Edward VII on his 66th birthday.[2] A slightly larger "black diamond" named Seigo, weighing 3167 carats or 0.633 kg, was found in Brazil in 1895. Black Diamonds or carbonados are not of gem quality and consist not only of diamond but also graphite and amorphous

---

[1] https://www.famousscientists.org/antoine-lavoisier/

[2] http://famousdiamonds.tripod.com/cullinandiamonds.html

© Springer International Publishing AG, part of Springer Nature 2018
I. Baker, *Fifty Materials That Make the World*,
https://doi.org/10.1007/978-3-319-78766-4_11

Fig. 11.1 The crystal
structure of diamond

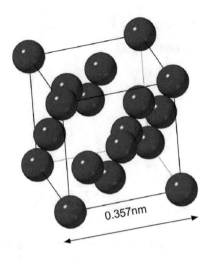

0.357nm

carbon.[3] While many individual large diamonds have names and are well known,
perhaps the best known diamond is the "Pink Panther", a fictitious diamond that
contains a flaw that looks like a leaping pink panther and is the basis of the Pink
Panther series of eleven movies.

Diamonds have some unusual properties. The word diamond comes from the
Greek *adámas* "unbreakable", and while diamonds are not truly unbreakable since
they are quite brittle and can be cleaved along specific crystal planes, they do have
the greatest hardness of any material. They also have an elastic modulus – a measure
of how easy it is to stretch a material - which is twice that of any other substance
1220 GPa versus 450–650 GPa for tungsten carbide, the next highest, and five times
the value for steel of 200 GPa.

Natural diamonds usually form over very long times (1–3 billion years) from
carbon-containing minerals under conditions of both high temperature and high
pressure that are found at depths within the Earth's mantle of 140–190 km. Diamonds
can also form when a meteorite impacts the Earth and transforms a graphite deposit
into diamonds, such as at the Popigai crater in Siberia, Russia. Impact diamonds are
not suitable as gems but are useful in industrial applications. It is also possible to
make diamonds on an industrial scale. Such synthetic diamonds can be made either
by reproducing the high pressure and high temperature (HPHT) conditions in the
Earth's mantle or at low pressure by chemical vapor deposition (CVD) in which a
gas such as 1% methane in hydrogen is heated to 700–1000 °C and energized. Thin
films of diamond over 15 cm in diameter can be produced by CVD. Both single
crystals and polycrystalline diamond can be produced using this technique. These
industrial diamonds are often flawed and colored. The first synthetic diamonds were
produced, using the HPHT route, in 1953 by the Swedish company ASEA, and
independently in 1954 by General Electric [1].

---

[3] https://www.costerdiamonds.com/famous-black-diamonds/

Fig. 11.2 Scanning electron microscope image of 9 micron industrial diamonds that are used for polishing metallographic samples

While natural diamonds used for jewelry can command high prices, most diamonds are used for industrial applications and worldwide almost 99% of these are synthetic diamonds, see Fig. 11.2. Only about 20% of natural diamonds are of gem quality but they have a market value of $72 billion in 2012.[4] The rest are used for industrial applications. Different outstanding properties of diamond lead to its use in different applications. Diamond thin films are used on some electronic devices to conduct away heat because diamond has the highest thermal conductivity of any material (up to ~ 4650 W/m.°C, which is over ten times larger than the 428 W/m.°C exhibited by the highest conductivity metal silver). At the same time, diamond has a very low electrical conductivity of 10^{-11} to 10^{-18} Siemens/m (copper is 6×10^7 Siemens/m), which is similar to some polymers such as polyethylene. Note that for most materials high electrical conductivity and high thermal conductivity go hand-in-hand. Industrial diamonds are used in grinding and polishing, in cutting tools, valve rings and other items that need high hardness and low wear rate. A particularly important application is the use on bits used for oil drilling, where the long life of such bits enables longer drilling before the drill has to be brought to the surface and replaced, which saves time and money.

Although alluvial diamonds have been known since ancient times, modern diamond mining spans only about 150 years, with the discovery of diamonds in Kimberley, South Africa in 1871. South African production dominated until the 1930s when it was overtaken by production in the Belgian Congo in the 1930s. The Belgian Congo production itself was later overtaken by production from a number of countries including Russia, Botswana, and Australia [2] with Russia being the largest current rough diamond producer at 18 million carats.[5] During the twentieth century the London-based company De Beers, which was founded by the Briton

---

[4] http://engagementrings.lovetoknow.com/wiki/What_Are_Diamonds_Used_For

[5] http://investingnews.com/daily/resource-investing/gem-investing/diamond-investing/top-industrial-diamond-producing-countries-drc-russia-australia/

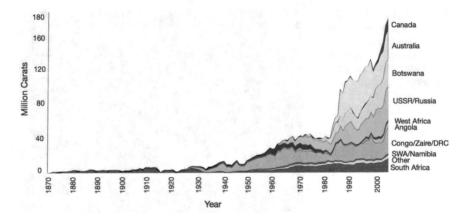

Fig. 11.3 Graph of global diamond production by carat weight from 1870 to 2005 for eight countries, one region in Africa, and all other producers. (After Ref. [2])

Cecil John Rhodes (1853–1902) and is now owned by the British-South African company Anglo American plc., had a monopoly over the diamond market [3] and still has a dominant position. The demand for natural diamonds has grown almost monotonically over time, but there was a dip in 2008 and global production has been fairly flat at around 120–130 million carats since then,[6] see Fig. 11.3. While mining takes place around the world, over 75% of polished diamonds are produced in India. In contrast, most synthetic diamonds are manufactured in China, which in 2013 produced 79% of the World supply with 4 billion carats (by comparison the U.S.A. produced only 125 million carats).[7]

Diamond has a lustrous future particularly if the cost of producing diamond films can be reduced. In that case applications like non-stick cookware could take advantage of the low friction, low wear rates, high thermal conductivity, resistance to corrosion and low coefficient of thermal expansion.

References

1. Amato, I. (1998). *Stuff: The materials the world is made of.* New York: Avon Books, Inc. ISBN-10: 0380731533.
2. (Bram) Janse, A. J. A. (2007, Summer). Global rough diamond production since 1870. *Gems and Gemology, 43*(2), 98–119.
3. Chang S.-Y., Heron, A., Kwon, J., Maxwell, G., Rocca, L., & Tarajan, O. (2002). The global diamond industry. *Chazen Web Journal of International Business.* The Trustees of Columbia University in the City of New York. https://www.gsb.columbia.edu/chazenjournal.

---

[6] http://www.bain.com/publications/articles/global-diamond-industry-report-2016.aspx

[7] http://investingnews.com/daily/resource-investing/gem-investing/diamond-investing/top-industrial-diamond-producing-countries-drc-russia-australia/

Chapter 12
Gallium Arsenide

You have probably never seen gallium arsenide (GaAs) and may not have even heard of it but every day you likely encounter devices that use this metallic compound and its related compounds aluminum gallium arsenide and indium gallium arsenide. Gallium arsenide has a similar crystal structure to silicon, but each atom of gallium has an arsenic atom nearest neighbor and vice versa, see Fig. 12.1. These materials are the core, along with the compound indium phosphide and its derived compounds (which are mostly used in telecommunications), of semiconductor lasers, which are also sometimes called semiconductor diodes or injection lasers. The function of a semiconductor laser is to turn electrical energy into light energy in a directed narrow beam. The best semiconductor lasers can turn around 70% of the input electrical power into light.[1] They are used in CD and DVD players, laser pointers, barcode readers, and laser printers and are also a key component in optical fiber data communications, see Fig. 12.2. In addition to being small enough to fit into these devices, they are both very reliable and long lasting. Semiconductor lasers typically operate in the wavelength range 0.4–2.1 microns (a micron is a millionth of a meter), that is from the violet end of the visible spectrum to the invisible infrared end of the spectrum [1] - note that the typical human eye can only see wavelengths of 390–700 nm.

Polycrystalline gallium arsenide is produced by reacting arsenic vapor with molten gallium in a sealed quartz ampoule at elevated temperature. Single crystals can be made in a number of ways. The most common is the Vertical Gradient Freeze process in which a crystal with a low-defect density is grown with a particular orientation from a seed by slowly freezing molten gallium arsenide by moving a temperature gradient along a sample. The single crystals are then sliced into wafers. Other related techniques for growing single crystal boules of GaAs include the Bridgman-Stockbarger technique and the Czochralski method. Thin films of GaAs can be produced directly using gaseous reaction techniques including Chemical Vapor Deposition (CVD) and Molecular Beam Epitaxy (MBE).

---

[1] https://www.photonics.com/EDU/Handbook.aspx?AID=25099

© Springer International Publishing AG, part of Springer Nature 2018
I. Baker, *Fifty Materials That Make the World*,
https://doi.org/10.1007/978-3-319-78766-4_12

Fig. 12.1 The crystal structure of gallium arsenide and aluminum gallium arsenide (zinc blende structure): the Ga atoms are depicted in blue and the arsenic atoms are red

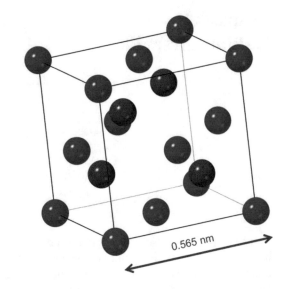

0.565 nm

Fig. 12.2 A laser pointer. The laser beam is red but so bright that it beyond the dynamic range of the camera used to photograph it

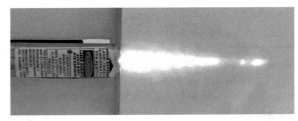

Robert N. Hall (1919–2016) of General Electric was granted the first patent for a semiconductor laser in 1962 and Nick Holonyak, Jr. (1928-) of IBM demonstrated the first visible light (red) laser diode in the same year. It wasn't until 1975 that Diode Laser Labs (NJ, USA) introduced the first commercial semiconductor laser. Five years later both Philips and Sony introduced CD players, which used semiconductor lasers, starting the huge growth in their use. Now 99% of lasers sold are semiconductor lasers.

The simplest concept of a semiconductor laser is shown in Fig. 12.3. Here, electrons are injected from an n-type semiconductor and holes are injected from a p-type semiconductor into an active layer or junction where they combine to produce light by stimulated emission, which emerges at one end of the active region and is focused by a lens. More typically, devices termed double heterostructure lasers, shown in Fig. 12.4, are used because these are more efficient. These will typically have GaAs contact layers that are joined to metal contacts and AlGaAs barrier layers with a thin GaAs active layer where the light is produced. The reflecting ends are produced by cleaving the resonator and the ends are coated with a dielectric so that one end fully reflects the light that is produced while the other end partially reflects the light and partly allows the light to escape. The typical output of a semiconductor laser is of the order of 1 mW. Laser pointers are less than 5 mW for eye safety since although

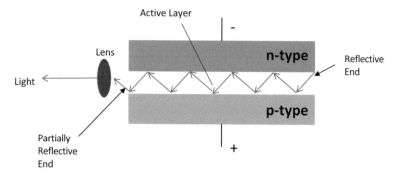

Fig. 12.3 Schematic of a simple heterostructure semiconductor laser. In practice the active layer is much thinner (at approximately 100 microns thick) than the other two regions

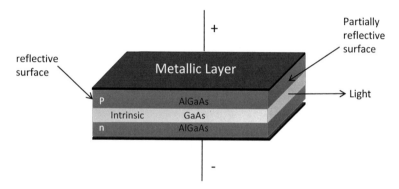

Fig. 12.4 Schematic of a Fabry-Perot double heterostructure semiconductor laser. The GaAs layer is about 200 nm thick. The AlGaAs layers are typically 1–2 microns thick

the power is low it is a highly-directional focused beam. By comparison an incandescent light bulb typically consumes 100 W of electricity but the visible light emission efficiency is only around 5%, and the radiation is in all directions so the power density (power per unit area) is quite low.[2] Although individual semiconductor lasers are low power they can be combined into bars and stacked to give over a hundred times more power – a 40 W bar operating at a voltage of 2 V would require 40–50 A to power it.

While gallium arsenide's main use is in semiconductor lasers, it also finds use in various semiconductor electronics because of the higher electron mobility and larger band gap than silicon. The higher electron mobility means that devices can be produced that work at frequencies greater than 250 GHz. The larger band gap also means that gallium arsenide devices are less sensitive to heat and radiation, which is useful for space electronics.

---

[2] Comment from Prof. Jifeng Liu, Dartmouth College.

Gallium arsenide solar cells are also more efficient (at >30% for a two junction cell), but more expensive, than silicon solar cells and absorb light more readily, again meaning that thinner layers can be used. This makes them useful for specialized applications where high efficiency and light weight are important, but cost is less important such as in Space applications.

However, for most electronics and photovoltaic applications silicon is the material of choice because it is much less expensive, easier to grow single crystals, and silicon oxide can be formed on the surface, which is an excellent insulator in electronic devices. Nevertheless gallium arsenide and its alloys have a bright future and no doubt new applications will be found.

Reference

1. Amato, I. (1998). *Stuff: The materials the world is made of.* New York: Avon Books, Inc. ISBN-10: 0380731533.

Chapter 13
Glass

Glass is a supercooled liquid in which the atoms do not have time during cooling from the melt to find their correct locations to form a crystal. The simplest glass, composed of the two elements that are most common in the Earth's crust, is pure silica or silicon dioxide (SiO_2), which melts at over 1650 °C. The resulting glass is called fused silica or quartz glass. It has low thermal expansion of 5.5×10^{-7} °C^{-1} (one third of that of common soda-lime glass), is very resistant to thermal shock, corrosion resistant, is hard, and is strong at high temperature. Thus, it is used in applications like chemical glassware and furnace tubes and crucibles. It is, of course more expensive than soda-lime glass because of the high temperature required to melt it.

Slower cooling of molten silica enables the atoms to crystallize either as cristo-balite, the equilibrium structure that melts at 1713 °C, or quartz, the form in equilibrium below 1050 °C, which is the second most common mineral in the Earth's continental crust. Both crystal structures are a bit complicated: cristobalite has a hexagonal crystal structure, while quartz adopts a lower symmetry trigonal crystal structure. Whether crystalline or amorphous, silica consists of repeat units of SiO_4^{4-}, see Fig. 13.1, which are joined together by so-called bridging oxygens. The difference between the crystalline state and the amorphous state is how these SiO_4^{4-} units stack together. Figure 13.2 shows the essential difference between crystalline silica and glass silica in 2-D, where it is evident that the glass has no long-range order, or to put it another way the structure does not repeat itself in a regular fashion. Glass, if bubble-free, is transparent. Glass is metastable, that is, it is not in the equilibrium state. If it crystallizes, a process referred to as devitrification, it becomes opaque since all the internal surfaces, called grain boundaries, of the crystals formed would scatter light.

The earliest glass used was naturally-occurring glass. Nature produces two glasses: obsidian, a black or blackish-green glass that is formed in volcanoes, and fulgurites, sometimes called petrified lightening, which are hollow or branched rods formed when lightening strikes sand, see Fig. 13.3. While its not clear what the later

© Springer International Publishing AG, part of Springer Nature 2018
I. Baker, *Fifty Materials That Make the World*,
https://doi.org/10.1007/978-3-319-78766-4_13

Fig. 13.1 Top: A SiO_4^{4-} tetrahedron; bottom: showing how the tetrahedral are joined together

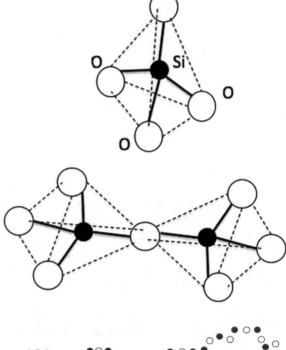

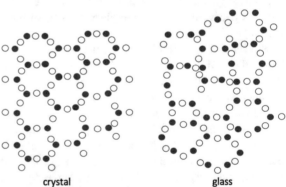

Fig. 13.2 Two-dimensional schematic showing the difference between crystalline and amorphous (non-crystalline) silica

could be used for, if anything, obsidian was valued in early societies because a very sharp cutting edge could be produced on pieces of obsidian.

The earliest use of manmade glass was circa 2500 B.C. as a glaze on ceramic bowls in northern Syria and Mesopotamia [1–3].[1] This ancient glass was made from silica and plant ash. The first glass vessels were produced in this region around 1500 B.C. Glass making emerged in South East Asia around the same time [1]. At some point, instead of plant ash, natron was added to silica to make glass. The key innovation in making glass was these additions of plant ash or natron. In the ancient

---

[1] http://www.historyofglass.com/

Fig. 13.3 Fulgurites, which are glass tubes are produced by lightening striking sand. (Courtesy of Steve Kirsche)

world, fires and furnaces could not reach the temperature required to melt silica, which is essentially sand. Ancient glasses were made either from ash derived from halophytic plants (plants that grow in highly saline water), which consists of soda (Na_2O) and lime (CaO), or natron, which is a natural substance that is mostly sodium carbonate with about 17% sodium bicarbonate and small amounts of sodium chloride and sodium sulfate. These were added to sand containing the remains of seashells, which are made of calcium carbonate that can be decomposed to lime and carbon dioxide upon heating. In these ancient glasses, which were typically of 65–70% silica, the soda addition significantly lowered the melting point to 1000-1100 °C, which made glass making possible. The addition of lime was necessary since glass made of only silica and soda dissolves in water, which the lime prevents. In some ancient glasses dating to the beginning of the first millennium B.C. potassium oxide was used in place of the lime. Interestingly 90% of modern glass is soda-lime glass, which is used in applications such as windows and bottles, that is not too different chemically from this ancient glass being ~74% SiO_2, 13–14% Na_2O, 10% CaO with smaller additions of MgO and Al_2O_3, and minor amounts of other metal oxides. The ratios of the oxides strongly affect the melting temperature with a minimum melting temperature at 725 °C for glass containing 73.1% Silica, 21.9% soda and 5.1% lime.

Glassmaking subsequently flourished along the eastern Mediterranean with products such as jewelry, cases and jugs. Someone in the first century B.C. found that glass could be blown, a practice that has been used ever since. This enabled glass to be made both faster and more cheaply. The Romans expanded the use of glass, making mirrors, square bottles that were blown in molds, and the first windows from small panes held together with lead. Glassmaking became common in the first century A.D. and by the twelfth century the making of stained glass windows for the growing number of cathedrals was at its peak [4]. By the thirteenth century glass was being used for eyeglasses and by the sixteenth century glass was being used to make lenses for telescopes.

Up to the sixteenth century, glass for windows was usually cut from large discs of crown glass, which was made by blowing a large bubble of glass and then spinning it around until it flattened out. This processing technique created glass that was thicker in the middle than on the outside. A vivid account of glass making in the late eighteenth century can be found in the novel *The Glass-Blowers* by Daphne Du Maurier. By around 1825 glass sheets for windows were mainly produced by blowing cylinders that were 2–2.5 m long by about 30 cm diameter, which were cut open, flattened, and then cut into panes. In 1848, the English inventor Sir Henry Bessemer (1813–1898) developed a process where molten glass could be poured between rollers, but the resulting solid glass had to be polished after processing, making this an expensive manufacturing method. A huge breakthrough was the Pilkington float glass process, invented in 1952 by Sir Alastair Pilkington (1920–1995). In this process molten glass is floated on bed of molten tin in an inert atmosphere (to prevent the tin from oxidizing), directly producing the window glass of the desired thickness with perfectly parallel surfaces without the need for further processing. The scrap glass or cullet produced during processing is fed back into the glassmaking reducing the melting temperature of the initial ingredients and reducing energy costs by about a fifth.[2] There are about 450 float glass plants around the world producing over a total of over a million tonnes of plate glass per year, which is 80% of the plate glass produced.

Since fused silica, which is pure silicon oxide makes such a good glass, why don't we make all glass out of pure silica? The answer is that pure silica's melting point is very high, and by adding other oxides, called network modifiers, like the Na_2O and CaO that we met earlier, we can lower the melting point considerably and make forming various products much easier and less expensive since high temperatures cost more energy. These network modifiers produce non-bridging oxygens that do not connect two SiO_4 tetrahedra, see Fig. 13.4. The addition of other oxides, such as boron trioxide (B_2O_3) and lead oxide (PbO) confer different properties. Although lead additions had been made to glass before Englishman George Ravenscroft (1618–1681) was awarded a patent in 1673 for his lead glass, which contained around 30% PbO, he was the first to perform industrial scale processing of clear lead crystal glassware. Lead glass is also useful for optical instruments, such as prisms, see Fig. 13.5. It is also used for radiation shielding and for fine crystal glass. Unfortunately, wine or port stored in the latter will leach the lead from the glass, with poor consequences for one's long-term health. Borosilicate glass (boron trioxide mixed in with the silica) was developed by the German Friedrich Otto Schott (1851–1935). This glass is much more resistant to thermal shock than soda-lime glass and has a thermal-expansion expansion coefficient one third of that of

---

[2] http://www.pilkington.com/en/us/architects/resource-library/float-glass-process-video

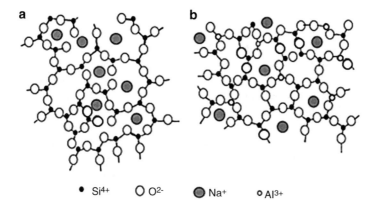

Fig. 13.4 Schematic showing how network modifying ions (Na⁺ and Al³⁺) produce non-bridging oxygen atoms that disrupt the silica network

Fig. 13.5 The cover of the Pink Floyd album shows a prism splitting light into its different colors. While artistic, the physics is incorrect since the splitting should also occur at the internal interface (arrowed) of the prism due the change in refractive index from air to glass

soda-lime glass (3.3×10^{-6} °C^{-1}). Glass can also be colored by adding different metallic salts (FeO or Cr_2O_3), or painting for both everyday use and for use for art objects.

Below the glass transition temperature (see Box), glass is brittle, typically fracturing purely elastically at a strength of 50 MPa or less in tension or flexure. It is particularly susceptible to scratches or notches which can concentrate the stress and cause premature fracture. On the other hand, thin hairs of glass can be very strong, since there are no defects. Glass fibers are used in glass fiber reinforced polymers (GFRP). These composites are discussed elsewhere in this book. In addition to changing the chemistry, one can change the properties by processing or making composites. Tempered glass is made by rapidly cooling the surface of the glass with air jets so that when the center of the glass sheet cools it produces compressive forces at the surface of the glass. Since glass is much stronger in compression (by a factor of up to 20), tempered glass is much tougher than other glass and when it fractures it breaks into small chunks rather than shards. It is used in windows for vehicles, shower doors, glass doors and tables, refrigerator shelves. Laminated glass is two sheets of glass that have a layer of polymer with the same refractive index sandwiched between them. When the glass breaks the glass is held in place. Thus, it is used in vehicle windshields. It has the added advantage that the polymer inter-layer stops UV radiation. Bulletproof glass is a variation on this where several lay-ers of glass and polycarbonate alternate - the more layers the more "bulletproof" and the thicker and heavier the glass.

While drinking containers and bowls could be made out of other materials than glass, many uses of glass are because it is transparent. Thus, until the relatively recent advent of transparent polymers, windows and eyeglasses had to be made out of glass. Other useful properties of glass is its durability and resistance to corrosion. Glass wasn't used as a structural material until Sir Joseph Paxton's (1803–1865) Crystal Palace, which was built for the Great Exhibition in London in 1851.

The amount of glass used continues to increase along with new types of glass, including photochromic, electrochromic, thermochromic, self-cleaning, low E-glass (reflects infrared radiation for energy savings in buildings) and newer uses such as for optical fibers, gorilla glass for touchscreens, and solar panels. The future of glass involves more recycling. Glass can be made from 95% recycled glass saving the acquiring of raw materials and energy in the processing. Currently, recycling rates for glass are lower than those for aluminum and some kinds of plastics, and stand at less than 40% in the U.S.A., providing much room for improvement.

Box

Materials Scientist use glass to have a broader meaning than just silica or alloys of silica. They use it to mean any material, which when rapidly cooled has no long-range structure. Thus, Materials Scientist's talk of metallic glasses, which are alloys that have been cooled rapidly to prevent crystal for-mation. A key difference between glasses based on silica and metallic glasses is that cooling rates on the order of 100,000 °C per second are typically need to form a metallic glass although some alloys, such those based on palladium-copper-tin, only need cooling rates of 100-1000 °C per second. Glasses differ from crystalline materials in many ways. One is in thermal expansion behav-ior. A crystal melts abruptly and has a large change in volume when it melts. In contrast the change in volume with temperature for a glass is gradual, see Fig. 13.A. Similarly, glasses exhibit a slow increase in viscosity as they cool. This slow change in viscosity and volume change is the reason that glasses can easily be blow molded and injection molded.

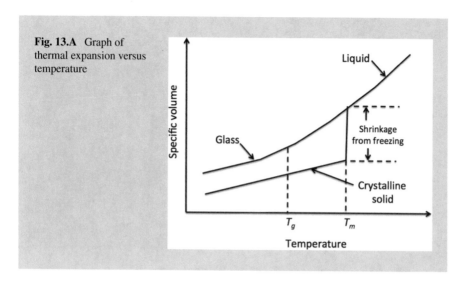

Fig. 13.A Graph of thermal expansion versus temperature

References

1. Whitehouse, D. (2002). The transition from Natron to plant ash in the Levant. *Journal of Glass Studies, 44*, 193–196.
2. Degryse, P. (Ed.). (2014). Glass making in the Greco-roman world: Results of the ARCHGLASS project. Leuven, Belgium: Leuven University Press. ISBN 978 94 6270 007 9.
3. Henderson, J. (2013). *Ancient glass*. Cambridge, England: Cambridge University Press. ISBN 978-1-107-00673-7.
4. Amato, I. (1998). *Stuff: The materials the world is made of*. New York: Avon Books, Inc.. ISBN-10: 0380731533.

Chapter 14
Glass Fiber Reinforced Polymers

Glass fiber reinforced polymer (GFRP), commonly called fiberglass, is used for a huge variety of applications including boat hulls, car bodies, roofing shingles, pipes, flooring and various containers and storage tanks. GFRP is referred to as a composite material since it is a combination of two different materials, that is, glass fibers in the form of either a woven fabric or a chopped mat in a polymer matrix. Most often used for the matrix are the thermosetting polymers (polymers which cannot be remelted) epoxy or polyester, and less-commonly the thermosetting polymers vinyl ester, phenolics and silicones, or the thermoplastic polymers (polymers which can be remelted) nylon, polycarbonate and polystyrene. The fibers are coated with a coupling agent to aid bonding to the polymer matrix and a sizing agent to protect the fibers [1], see Fig. 14.1.

Fiberglass is by far the most common fiber composite for non-aerospace applications. Compared to other fibers incorporated into polymer composites, glass fibers have a tensile strength (3400 MPa) comparable to that of aramid fibers such as Kevlar (3400 MPa) but much lower than that of carbon fibers (6350 MPa). The elastic modulus (the ability to resist stretching) of glass fibers is low (72 GPa) compared to that of aramid fibers (186 GPa) and very low compared to that of carbon fibers (400 GPa). GFRPs are also less brittle than carbon fiber composites. Kevlar and carbon fiber polymer composites are more expensive by a factor of 3–10 than GFRPs. However, GRFPs themselves are 4–6 times more expensive than plain carbon steel and more than twice as expensive as wood.

Flawless continuous glass fibers thousands of meters long can be easily extruded from molten glass through 0.005–0.025 mm diameter bushings over a thousand at a time at a speed of several thousand meters per minute and then collected into bundles called tows or rovings.[1] Several different compositions are used to make continuous glass fibers, which consequently have different properties and different costs and are used for specific applications. The most commonly used (~90%) is electrical grade glass or, more succinctly, E-glass so named because it is an electrical

---

[1] http://www.britglass.org.uk/glass-fibre-manufacture

© Springer International Publishing AG, part of Springer Nature 2018
I. Baker, *Fifty Materials That Make the World*,
https://doi.org/10.1007/978-3-319-78766-4_14

Fig. 14.1 A glassfiber
swatch

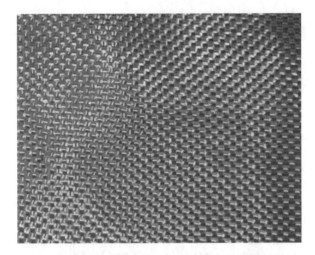

Fig. 14.2 Glass fiber-
epoxy composite plate

insulator and useful in applications where transparency to radio signals is a requirement. E-glass, an alumino-borosilicate glass, which is typically (by weight) 54%
SiO_2, 14 Al_2O_3, 17% CaO, 5% MgO, 10% B_2O_3 with less than 2% $Na_2O + K_2O$,[2] has
a wide range of desirable properties including low cost, high tensile strength
(3400 MPa), high elastic modulus (72 GPa), low density (2550 kg/m^3), and, hence,
a high specific strength (strength-to-weight ratio) of 1300 $kN \cdot m \cdot kg^{-1}$ (compared
to 211 $kN \cdot m \cdot kg^{-1}$ for aluminum), relatively low melting point (850 °C), good
chemical resistance, and it is relatively insensitive to water. S-glass fibers, which
were developed for military applications such as missile motor casings and made
with magnesium aluminosilicates (65% SiO_2, 25% Al_2O_3, 10% MgO), are of higher
strength (4700 MPa) and elastic modulus (89 GPa) than E-glass fibers, and also
have better high temperature properties and corrosion resistance. C-glass, which has
similar mechanical properties to E-glass, is made with calcium borosilicates (67%
SiO_2, 4% Al_2O_3, 13% CaO, 3% MgO, 8% Na_2O, 5% B_2O_3) [2] for improved chemical resistance and is useful in environments containing acids.[3]

Glass fibers have high tensile strength, but a long thin fiber (of any material) will
buckle in compression producing a low strength. The surface of fibers can also be
damaged introducing cracks. Hence, incorporating the fibers in a polymer matrix
improves the compressive properties and protects the fibers, see Fig. 14.2. In a

---

[2] http://www.azom.com/article.aspx?ArticleID=764

[3] http://www.build-on-prince.com/glass-fiber.html#sthash.DayHwLvA.dpbs

continuous fiber composite, the properties of the fibers and matrix are averaged. For example, incorporating E-glass fibers with the properties noted above in a polyester resin matrix which has a tensile strength of 40–90 MPa and an elastic modulus of 2.0–4.4 GPa, produces a typical composite with a tensile strength of 100–344 MPa and an elastic modulus of 5.5–31 GPa [3], the exact values depend on the volume fraction of fibers and their arrangement. Whereas polyester resins are cheaper and have better corrosion resistance, an epoxy matrix offers higher mechanical performance. GFRP composites are limited to low temperature use since the polymer matrix will start to soften around 200 °C. In addition to use in polymer matrix composites, glass fibers can also be used to reinforce rubber, cement, asphalt and gypsum.

Short glass fibers are produced by melt spinning in which a crucible containing molten soda lime glass with several hundred holes is spun around. The fibers could be made from pure silica (SiO_2), but this melts at very high temperature: the addition of lime (CaO) lowers the melting point significantly. The melt spinning process was invented in 1933 by the American inventor and engineer Russell Games Slayter (1896–1964) while working for Owens-Illinois Glass Co., Toledo, Ohio. Such short fibers are used to make glass wool, which is used as thermal insulation in buildings.

How to make thin glass fibers by melting and stretching glass was known by Phoenicians, Egyptians and Greeks[4] and they occur naturally as Pele's hair - named after the Hawaiian goddess of volcanoes - which is formed in volcanoes from molten basaltic glass.[5] Pele's hair is yellow and can look like human hair and is so light that it can be carried by the wind several kilometers from a volcano.

In 1893, Georgie Eva Cayvan (1857–1906) a popular stage actress was the first person to wear a glass dress in the play "American Abroad".[6] The dress was made by the Libbey Glass Company, which was started by the American entrepreneur Edward Drummond Libbey (1854–1925) and on moving the company to Toledo, Ohio in 1888 made that city the glass capital of the world in the early twentieth century. High fashion dresses have been made of glass or have incorporated glass to this day. They are, of course, not very practical. However, glass fibers are incorporated into textiles.

The trend with glass fiber is for increased performance by developing new compositions at ever lower cost, with China leading the charge in the latter trend. In 2011, the global production of glass fibers was 4.4 million tonnes with China producing an astonishing 76% of global output with 43% of Chinese output taking place in only three companies.[7] China is also the largest consumer of glass fiber.

---

[4] http://www.compositesworld.com/articles/the-making-of-glass-fiber

[5] http://www.amusingplanet.com/2015/09/peles-hair-and-peles-tears.html

[6] http://columbus.iit.edu/bookfair/ch24.html#881

[7] http://www.prnewswire.com/news-releases/global-and-china-glass-fiber-industry-report-2012-2015-185273212.html

References

1. Strong, A. B. (2000). *Plastics: Materials and processing*. Upper Saddle River: Prentice Hall. ISBN: 0-13-021626-7.
2. Flinn, R. A., & Trojan, P. K. (1990). *Engineering alloys and their applications*. Boston: Houghton Mifflin Company. ISBN: 0-395-43305-3.
3. Smith, W. F. (1990). *Principles of materials science and engineering*. New York: Mcgraw Hill Publishing Company. ISBN-13: 978-0070592414.

Chapter 15
Gold

Transmutation of a base metal into gold (symbol Au from the Latin Aurum) is possible not through the alchemist's chemistry, but by nuclear transmutation as demonstrated by the American Glenn Seaborg (1912–1999) and his collaborators who turned bismuth into gold using a particle accelerator at the Lawrence Berkeley Laboratory in California [1] in 1981. Turning lead into gold by such an approach is difficult and definitely not a profitable venture - the reverse nuclear transmutation is easier but even less valuable.

Gold can be obtained by sluicing or panning of alluvial ores, which directly yields small pieces of gold. This is only practiced currently by craft workers. Gold is mostly produced by crushing low-grade ores and then leaching them using cyanide. Each gram of gold produced this way results in about a tonne of waste ore [2] through processing that uses large quantities of water and energy. On the other hand, the amount of gold from recycling is as large as the gold from mining: why would anyone throw gold away given its high value? A large source of recycled gold is mobile phones.

Gold is mined in many countries throughout the world. In 2017, China was the largest producer at 440 tonnes, with Australia, Russia and the U.S.A. as the next largest producers at 300, 255 and 245 tonnes, respectively.[1] World production was 3150 tonnes. Estimated World reserves of gold are 54,000 tonnes, but it is worth noting that the Oceans contain 10 million tonnes of gold. Unfortunately, the concentration of gold in seawater is only 10 parts per trillion, and no-one has found an economical way to extract this gold.

Because of its scarcity - all the refined gold in the world through 2014 only amounts to a cube 21 m on each side or 183,600 tonnes.[2] Gold has been valued by most societies since the dawn of humanity being used as jewelry and as a store of wealth. Its value is related to its scarcity.

---

[1] USGS, Mineral Commodity Summary, Gold. 2018.

[2] http://www.gold.org/supply-and-demand/supply

© Springer International Publishing AG, part of Springer Nature 2018
I. Baker, *Fifty Materials That Make the World*,
https://doi.org/10.1007/978-3-319-78766-4_15

Fig. 15.1 The face
centered cubic structure
adopted by gold. The side
of the unit cell is
0.4079 nm

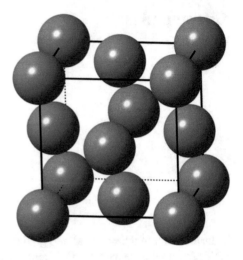

Gold is one of four colored metals, the others being copper, cesium and osmium, and one of three metals that are commonly found naturally in the metallic state and not tied up in compound – the other two metals commonly found in nature are copper and iron, with both the gold and iron arriving in meteorites. While pure gold is a bright yellow color, gold can take on other colors when it is alloyed with various other metals such as: white gold (alloyed with nickel, manganese or palladium); blue gold (alloyed with iron); green gold (alloyed with silver), a naturally-occurring alloy known to the Lydians, who ruled in western Anatolia in the Late Bronze Age, as electrum; red gold (alloyed with copper); and rose and pink gold (alloyed with copper and silver).

Gold, which has a f.c.c. crystal structure (see Fig. 15.1), is very soft and ductile in its pure form. Its ductility means it can be beaten into very thin sheets, so thin that they are optically transparent. This malleability and its scarcity meant that gold has often found use as coins, starting in Lydia around 600 B.C.[3] The possession of large quantities of gold and its use in coinage has sometimes been a mixed blessing. The Roman's debasement of their gold and, particularly, silver coins starting with the Emperor Nero (37–68 AD) rather than solving the problem of a lack of funds led to inflation.[4] And the vast wealth of gold and silver obtained by the Spanish Empire (1492–1898) meant that Spain developed little industry [3]: the bullion caused inflation, contributing to the fall from the richest country in the world.

Even when not used as coins, gold has been used to back the value of circulating currency. The United Kingdom adopted the Gold Standard in 1717, the United States in 1819 and many other countries in the 1870s. The Gold Standard was such that a specific amount of gold was fixed to a certain value of the currency, and the government guaranteed that the currency could be converted into gold. The adoption

---

[3] http://rg.ancients.info/lion/article.html

[4] http://dirtyoldcoins.com/Roman-Coins-Blog/?p=190

Fig. 15.2 One of the
major uses for gold is in
jewelry

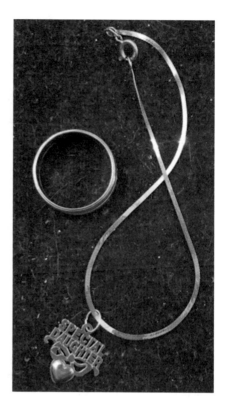

of the Gold Standard by many countries meant that the currency exchange rates were also fixed. Eventually, all countries abandoned the Gold Standard, with Switzerland being the last country to do so in 1999. But gold is still purchased as a hedge against turmoil in the World's economies.

At atomic number 79, Gold is one of the heaviest metals with a density of 19,300 kgm^{-3}. Attempts have been made to insert rods of tungsten, which has an almost identical density of 19,250 kgm^{-3}, into gold bars in order to cheat the buyer.[5] The presence of such tungsten rods can be determined without slicing open the gold bars by ultrasonic testing, and there is commercial equipment that can serves that purpose.

If asked which is heavier an ounce of tungsten or an ounce of feathers, you would probably say they weigh the same, and you would be right. If asked which is heavier an ounce of tungsten or an ounce of gold, the answer is an ounce of gold since gold is measured in troy ounces, which are roughly 10% greater than avoirdupois ounces that is used to measure most materials.

Gold is very soft and when used in pure form in jewelry, which amounts to 50% of its current use, it is very easy to deform and scratch. Pure gold is often called 24 carat gold, that is, 24 parts out of 24 are gold. The gold in jewelry often contains

---

[5] http://www.businessinsider.com/tungsten-filled-gold-bars-found-in-new-york-2012-9

Fig. 15.3 Iron pyrite
(FeS_2), sometimes known
as Fool's gold

either silver or copper, see Fig. 15.2. These foreign atoms substitute for gold atoms in the lattice and harden it – by making linear defects called dislocations more difficult to move. For instance, 18 carat gold, which is 18 parts out of 24 gold or 75% gold, is about 40% harder than pure gold [4].

One should always keep in mind that "all that glitters (or glistens) is not gold". Small pieces of pyrite or iron pyrite (FeS_2) can be mistaken for gold particularly when panning for gold, a practice started by the Ancient Egyptians. Hence it's nickname "fool's gold". However, the resemblance is superficial since pyrite is harder, much more brittle and usually occurs as crystals with clear facets (Fig. 15.3).

Apart from as a store of wealth and jewelry, until recently gold has had little practical use. Nowadays, gold is used extensively for contacts and wire bonding in electronics both because it has one of the highest electrical conductivities amongst metals of 44×10^6 Siemens/m (only silver and copper have higher electrical conductivities – by comparison a typical steel only has an electrical conductivity of 1×10^6 Siemens/m) and because it is malleable, non-toxic and relatively inert and, thus, resistant to oxidation and corrosion. However, even gold can have problems in electronic devices. One problem arises when gold bonds are made to aluminum. When heated to the normal operating temperature of a silicon chip, gold and aluminum films slowly diffuse into each other and form a number of gold-aluminum intermetallic compounds. The most notorious of these compounds is $AuAl_2$, which is called the purple plague in the semiconductor industry because of its color and because it is associated with bond failures. Aluminum and gold both diffuse into each other and in gold-aluminum compounds at different rates. The latter leads to a void formation, which can also be a source of failure [5], see Fig. 15.4.

Gold is also being used in various biomedical applications because of its inertness and lack of toxicity in the body. One such application is gold nanoshells, which consist of a dielectric core, typically silica, covered in gold producing a sphere typically 10–200 nm (a nanometer is a billionth of a meter) in diameter. These nanoshells, whose diameters are below the wavelength of visible light, absorb light of different colors depending on the size of the nanoshell with smaller shells absorbing longer

Fig. 15.4 Scanning electron microscope image showing voids forming (arrowed) in a gold bond to an aluminum film (darker layer). (From Ref. [5])

wavelengths. Thus, different sized nanoshells have different colors. The nanoshells can be injected into a cancerous tumor and heated with near infrared radiation to therapeutic temperatures [6].

In addition to electronics, jewelry and medicine, gold is also used for connectors, as a reflector of infrared and visible light and in specialized applications for heat shielding, such as on astronaut helmet faceplates [2].

Gold has a bright and shiny future, marred perhaps only by the problem that gold is one of the rarest elements (rated seventy-second in abundance) in the Earth's crust. However, even this is not a big problem since most gold is currently used for non-critical purposes, around 50% is used for jewelry, often as a method of wealth storage, and most gold is eventually recycled even from the small amounts in the many electrical connectors in your cell phone.

References

1. Aleklett, K., Morrissey, D., Loveland, W., McGaughey, P., & Seaborg, G. (1981). Energy dependence of [209]Bi fragmentation in relativistic nuclear collisions. *Physical Review C, 23,* 1044.
2. Ploszajski, A. (2016) Material of the Month: Gold, Materials World Magazine. pp. 58–60. IOM3.
3. Chaline, E. (2012). *Gold, fifty minerals that changed the course of history.* Buffalo: Firefly Books. ISBN: 13: 978-1-55407-984-1.
4. Suzuki, T., Vinogradov, A., & Hashimoto, S. (2004). Strength enhancement and deformation behavior of gold after Equal-Channel angular pressing. *Materials Transactions, 45,* 2200–2208.
5. Maiocco, L., Smyers, D., Kadiyala, S., & Baker, I. (1990). Intermetallic and void formation in gold Wirebonds to aluminum films. *Materials Characterization, 24,* 293–309.
6. Erickson, T. A., & Tunnell, J. W. (2010). Gold Nanoshells in biomedical applications. In *Nanotechnologies for the life sciences.* Chapter 1. Hoboken, NJ: Wiley. ISBN: 9783527610419.

Chapter 16
Graphite

Carbon exists in many forms: as the crystalline allotropes graphite and diamond, as amorphous carbon, and as the nanomaterials carbon nanotubes, graphene and Buckminsterfullerene that was named after Buckminster Fuller (1895–1983) and more often called buckyballs or fullerenes. The last three forms are one-dimensional, two-dimensional and three-dimensional nanostructures that were first synthesized in 1991, 2004 and 1985, respectively. The British chemist Harold Kroto (1939–2016) and the two American chemists Robert Curl (1933-) and Richard Smalley (1943–2005) were awarded the 1996 Nobel Prize in Chemistry for their discovery of buckyballs, while two British physicists Andre Geim and Konstantin Novoselov were awarded the 2010 Nobel Prize in Physics for their isolation of graphene. Carbon nanotubes were first synthesized by Japanese researcher Sumio Iijima (1939-) [1], but may have been observed as early as 1951 by Russian researchers L.M. Radushkevich and V.M. Lukyanovich [2]. Only graphite, which is found naturally as a gray metallic-looking mineral, is the stable or equilibrium form of carbon.

Graphite exists in two crystalline forms one hexagonal and one rhombohedral. Hexagonal (alpha) graphite consists of one atom thick layers of graphene in which carbon atoms spaced only 0.142 nm (nm is one billionth of a meter) apart are strongly bonded to each other in a hexagonal ring arrangement. Carbon's four bonding electrons are all utilized for bonding within the graphene layer so that the graphene layers themselves are only weakly held together by Van der Waals bonds and the carbon atoms are a much greater distance apart (0.335 nm), see Fig. 16.1. The rhombohedral (beta) form is very similar to the hexagonal form but the layers of graphene are stacked slightly differently, see Fig. 16.2. Graphite, which occurs naturally in igneous and metamorphic rock and in meteorites, is a mixture of these two crystal structures with the beta form accounting for up to 30% of the total.

The layered structure of graphite gives it some unusual properties. The sheets can be easily cleaved imparting graphite with a low room temperature tensile strength of only 14–69 MPa (compared to 450 MPa for a high strength precipitation-hardened aluminum alloy) and low coefficient of friction of 0.1 (compared to steel

© Springer International Publishing AG, part of Springer Nature 2018
I. Baker, *Fifty Materials That Make the World*,
https://doi.org/10.1007/978-3-319-78766-4_16

Fig. 16.1 The hexagonal crystal structure that is commonly adopted by graphite. The structure consists of layers of carbon atoms in hexagonal rings in which the distance between the atoms is 0.142 nm: the distance between the layers, which are only weakly connected together is 0.335 nm. The black box shows the hexagonal unit cell, which has axes in the horizontal plane of 0.247 nm, while the vertical axis is 0.679 nm

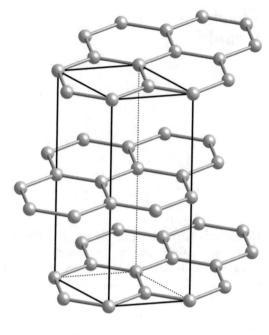

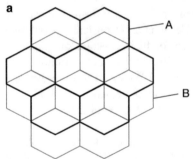

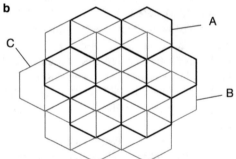

Fig. 16.2 Schematic viewing perpendicular to the rings of carbon atoms in (**a**) hexagonal, and (**b**) rhombohedral graphite. In hexagonal graphite the rings are stacked in an ABABAB sequence in which every second layer is in the same position as the first layer, see Fig. 16.1. In rhombohedral graphite the rings are stacked in an ABCABC sequence in which every third layer is in the same position as the first layer. Natural graphite is a mixture of these two crystal structures with the rhombohedral crystal structure accounting for up to 30% of the total

of 0.5–0.8).[1] Thus graphite is used as a lubricant in industrial applications.[2] The different atomic structures in the plane of the graphene sheets and perpendicular to them mean that the thermal, acoustic, mechanical and electrical properties are quite different within the graphene plane and perpendicular to it. This behavior is referred

---

[1] http://www.engineeringtoolbox.com/friction-coefficients-d_778.html

[2] https://hypertextbook.com/facts/2004/NinaChen.shtml

Fig. 16.3 The "lead" in a pencil is in fact graphite mixed with clay

to as anisotropy. One can imagine three of carbon's valence electrons bonding to nearby atoms and the fourth electron as being free to move in the graphene plane or being "delocalized". These delocalized electrons enable excellent electrical conductivity in the graphene plane, while the electrons cannot move out of the plane producing little electrical conduction perpendicular to the plane. However, when graphite is powdered, producing crystals of all orientations in the powder, the graphite conducts well in all directions. The thermal stability, good electrical conductivity ($\sim 1 \times 10^5$ Siemens/m compared to 6.2×10^7 Siemens/m for silver and 1×10^{-13} Siemens/m for nylon) and very high thermal conductivity of graphite (140 W/m · °C compared to 430 W/m · °C for silver and 0.24 W/m · °C for nylon) leads to its two major uses as refractories and electrodes in the high temperature processing of materials, and for electrodes in electric motors, batteries and fuel cells. It is also used in furnace windings, extrusion dies, as a lubricant and as a source of carbon in steelmaking.

Graphite mixed with clay is the "lead" in your pencil, see Fig. 16.3. The confusion over lead and graphite in pencils arises because the dark gray, metallic-looking graphite was originally called "black lead" or "plumbago" (the Latin for lead is *plumbum*) because it looks similar to the lead-containing ore galena. When you write with a pencil you are breaking the Van der Waals bonds. In fact the name graphite arises from its ability to mark paper from the German "Graphit," given by German mineralogist Abraham Gottlob Werner (1750–1817) based on the Greek word for "write" *graphein*.[2] About 7% of graphite use is for making pencils.

Nowadays, a small but important use of graphite is in some nuclear reactors, where it is used as neutron moderator because of its low neutron cross section. It has been used for the application since the first nuclear reactor was built, Chicago Pile-1, at the University of Chicago in 1942. Two major accidents have involved graphite fires in nuclear reactors, one in 1957 at Windscale, England and one in 1986 at Chernobyl, Ukraine. Both spread radioactive material across Europe.[3]

Graphite is historically important. The discovery of a huge deposit of graphite at Seathwaite, Cumbria, England in the mid-sixteenth century enabled the English to use it to line molds for casting musket balls and cannon balls. The result was more spherical cannon balls that could be fired both more accurately and farther and, thus, aided in beating off the Spanish Armada that attempted to invade England in 1588.

In carbon or graphite fibers the hexagonal structured planes of carbon atoms in graphite are aligned along the fiber axis, see Fig. 16.4. While amorphous carbon fibers may have been produced in the nineteenth century for use in the first practical light bulb filaments by Thomas Edison (1847–1931), the first high performance

---

[3] https://www.britannica.com/technology/nuclear-reactor/Liquid-metal-reactors#ref155206

Fig. 16.4 The structure of carbon fibers. The fibers are made from a polymer, such as PAN, by removing the hydrogen by heating at 400-600 °C and removing most of the nitrogen by heating to 600- 1300 °C. After the final step, nitrogen atoms remain at the edge of the fibers

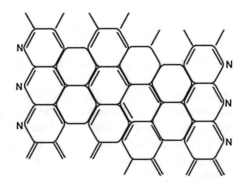

carbon fibers were produced at Union Carbide, near Cleveland, OH by Dr. Roger Bacon (1926–2007).[4] These fibers were not terribly strong since they contained only about 20% carbon, but improvements over the years eventually led to fibers that contain 99% carbon (the rest being mostly nitrogen) and to very high strength. Graphite fibers, which are 0.005–0.01 mm in diameter, are produced by taking organic fibers such as rayon, which are synthetic fibers of nearly pure cellulose made from wood pulp, or polacrylonitrile (PAN) fibers and carbonizing them by heating to 1000–3000 ° C in an oxygen-free environment. Since the resulting carbon fibers are easily damaged they are collected into tows consisting of thousands of fibers and protected with an organic coating, which are then woven into unidirectional fiber sheets. The fiber sheets are then usually layered so that the fibers are either at ±60° or parallel to the original fibers.[5]

The combination of both high elastic modulus or resistance to stretching (400 GPa versus 70 GPa for aluminum) and high tensile strength (up to 6350 MPa versus 570 MPa for high-strength aluminum alloys) with low density (1800 kg · m^{-3} versus 2700 kg · m^{-3} for aluminum) means that carbon fibers have very high specific modulus (density-compensated modulus) and high specific strength (density-compensated strength). The specific strength is 3500 kN · m · kg^{-1} versus only 211 kN · m · kg^{-1} for aluminum. This along with the low thermal expansion (-0.5×10^{-6} °C^{-1} in the fiber direction, i.e. contraction, and 7–10 × 10^{-6} °C^{-1} in the transverse direction versus ~ 23 × 10^{-6} °C^{-1} for aluminum), and high resistance to chemical attack provides a material for a variety of demanding structural applications. However, carbon fibers are normally made into a composite with the fibers aligned in one direction either with a polymer matrix (typically epoxy) or a graphite matrix. The latter are so-called reinforced carbon-carbon composites, which are useful for applications at temperature up to 2000 °C or even higher if it has an oxidation-resistant coating. Such composites are only used for applications where nothing else will work as well since they are very expensive such as: the leading edges of wings,

---

[4] https://www.acs.org/content/acs/en/education/whatischemistry/landmarks/carbonfibers. html#first-carbon-fibers

[5] http://zoltek.com/carbonfiber/

Fig. 16.5 Carbon fiber composite racquets are strong and light but prone to brittle failure as demonstrated here by one of the author's squash racquets. Some fibers are arrowed

the nose and of the space shuttle, the nose cones of missiles, and brakes on aircraft and Formula One racing cars.

In comparison with other typical reinforcing fibers, carbon fibers have the highest elastic modulus (400 GPa versus 72 GPa for glass fibers and 186 GPa for aramid fibers) and highest strength (6350 MPa versus approximately 3400 MPa for both glass fibers and aramid fibers).

Compared to the carbon fibers themselves, a carbon fiber-epoxy composite is not as strong and has a lower elastic modulus (the resistance to being deformed elastically) - it would have a slightly lower density compared to the fibers alone since epoxy has a lower density. For example, a composite that contains 60% fibers would have a strength of 760 MPa along the fiber axis and only 28 MPa perpendicular to this versus 6350 MPa for the fibers themselves. Similarly, the elastic modulus along the fiber direction, which is simply the weighted average of the elastic moduli of the matrix and the fibers, would be 220 GPa and only 6.9 GPa perpendicular to the fibers versus 400 GPa for the fibers themselves.

As the carbon fibers are normally woven into cloths, different orientations of the fibers in the cloths can be used to strengthen the material in the direction where the highest strength is most needed.

It seems that all new strong structural materials, and carbon fiber composites were no exception, start their life in golf clubs – golfers will pay anything to drive a ball a few more meters - and then work their way down into various sports racquets and into other sports, e.g. hockey sticks, bicycle frames, see Fig. 16.5. Even Oscar Pistorius, the South African sprinter, uses carbon fiber blades to replace his amputated legs.

The main use for carbon-fiber composites is in the aerospace industry, blades for wind turbines and in high-end racing cars. Carbon fiber/epoxy composites have been used for civilian aircraft since the mid-1970s. They are accounting for increasing amounts of such aircraft. In some of the newest jetliners such as the Airbus A350 XWB and the Boeing 787 Dreamliner both the wings and fuselage are made of carbon fiber composites: the A350 and Dreamliner have 53% and 40% composites parts, respectively. Using carbon fiber composites for such applications provides a number of significant advantages over traditional materials like aluminum. The most obvious is that there are significant weight savings, possibly up to 20% of the weight of the aircraft. It has been estimated that each kilogram reduction in weight produces a savings of $1 M over the lifetime of the plane.[6] More visibly evident are the new fuel-efficient designs made possible using carbon fiber composites such as the use of sweeping wing tips, which can produce 5% reductions in fuel consumption - more radical more fuel-efficient designs will follow in the future. Less evident, the use of composites also reduces the six million parts that go into a typical jetliner. For example, the Dreamliner uses one-piece barrel sections made of composites rather than multiple aluminum parts connected with 40,000–50,000 fasteners for the fuselage.[7] This reduces production costs and, presumably, maintenance costs. One effect of replacing aluminum with carbon-fiber/epoxy composites is that the aircraft skin is no longer electrically conductive. Thus, lightening conduction paths have to be built into composite aircraft. While these are huge advances, in 2016, Diamond Aircraft, an Austrian company, produced the world's first all-carbon fiber commercial aircraft a small trainer with a maximum take-off weight of 2130 kg.[8]

Virgin Galatic, a company owned by British business magnate Richard Branson (1950-), first flew an all-carbon fiber aircraft - even the flight control cables were made from carbon fiber - WhiteKnightTwo in 2008 that was built by American aerospace engineer Burt Rutan's (1943-) Scaled Composites company. At the time, this twin-fuselage aircraft had the largest aviation component made out of carbon fiber, the 42.7 m long wing [9] that will carry SpaceShipTwo under it's wing to around 15,000 m before releasing it and its paying customers into a sub-orbital flight.[10,11] Other uses are envisioned for the aircraft such as a forest-fire water-bomber.

While carbon fiber usage will continue to grow for aircraft, sporting goods and motor vehicles, usage is also slowly expanding from racing cars to more expensive automobiles where the high strength-to-weight ratio can be utilized. Carbon fiber composites are increasingly being used both to repair corroded bridges, and for new bridge building. Arches made of carbon fiber composite tubes can be taken flat to a site, inflated and filled on site with concrete. When the concrete is set, the arches are lifted into place as bridge supports. Then either a carbon fiber composite or concrete decking can be used. This provides rapid and easy construction of small bridges resulting in low-maintenance, long-lasting bridges that can largely be assembled without the need for heavy equipment and are cheaper,[12] see Fig. 16.6. The first public highway fiber composite bridge in Europe was the 10 m span West Mill Bridge over the River Cole in Oxfordshire, England built in 2002, which consisted of a carbon and glass fiber composite deck and longitudinal beams over the river that was one quarter the weight of the equivalent concrete and steel bridge. The lightness of the composite bridges means that they are easier to install in remote

[6] http://www.bbc.com/news/business-25833264

[7] https://www.rdmag.com/article/2006/11/dream-composites

[8] https://aviationvoice.com/worlds-first-all-carbon-fiber-aircraft-takes-first-flight-201605230951/

[9] https://www.dezeen.com/2008/08/03/virgin-galactic-unveils-whiteknighttwo/

[10] https://www.space.com/4869-virgin-galactic-unveils-suborbital-spaceliner-design.html

[11] https://www.space.com/5468-virgin-galactic-spaceline-mega-mothership-set-rollout-debut.html

[12] http://www.aitbridges.com/advantages/

Fig. 16.6 The Wanzer Brook Bridge in Fairfield, Vermont, U.S.A. under construction. The bridge was constructed from concrete-filled carbon-fiber composite tubes and has a carbon-fiber composite deck. This building technique is sometimes referred to as "bridge in a backpack", because of the lack of need for heavy equipment. Courtesy of the Vermont Agency of Transportation

locations. They also have the advantage that unlike steel and concrete the bridges are not corroded by de-icing chemicals. New uses of carbon fiber composites for structural applications seem only to be limited by the imagination.

One downside of this growth is that carbon fiber composites are not easily recyclable and do not degrade if thrown into landfills. While composites can be cut up and re-used, they do not have the strength of the original material.[13]

Graphite is obtained using both underground and open pit mining with China accounting for 708,000 tonnes or 60% of World Production in 2016. Graphite use will continue to grow, particularly as graphite fibers. The carbon nanomaterials, carbon nanotubes, graphene and buckyballs, have only recently been isolated and, thus, their uses are quite limited. However, they may be the future of carbon since they have some extraordinary mechanical, electrical and optical properties.

References

1. Lijima, S. (1991). Helical microtubules of graphite carbon. *Nature, 354*, 56–58.
2. Radushkevich, L. M., & Lukyanovich, V. M. (1952). O strukture ugleroda, obrazujucegosja pri termiceskom razlozenii okisi ugleroda na zeleznom kontakte. *Zurn Fisic Chim, 26*, 88–95.

---

[13] http://www.redorbFit.com/news/science/1112493049/the-dirty-secret-of-carbon-fiber/

Chapter 17
Gutta Percha

Gutta Percha has few uses nowadays apart from in endodontics, for which it has been used for over 150 years [1]. Gutta Percha, or technically trans-1,4-polyisoprene, is chemically identical to latex or natural rubber, which is cis-1,4-polyisoprene, see Fig. 17.1. Such molecules that are chemically the same but differ in the spatial arrangements of the atoms are referred to as stereoisomers. This difference is important since latex is amorphous (non-crystalline) whereas Gutta Percha crystallizes when it solidifies. While latex is produced from the sap of a rubber tree, Gutta Percha is a milky latex-like substance produced from a tree whose Malay name is Getah Perca. The Malay people had used Gutta Percha for many centuries before its introduction to the West in applications such as handles for knives and walking sticks amongst other uses.

Gutta Percha was first introduced to the West in 1656 by the English botanist and gardener John Tradescant the Younger (1608–1662) after traveling in the Far East. He gave it the name "Mazer Wood".[1] Unfortunately, the material was ahead of its time with no obvious applications and was simply a curiosity. It was re-introduced to the West by Scotsman Dr. William Montgomerie, Assistant Surgeon to the President in Singapore, for which he received the gold medal of the Society of Arts, London in 1844.[2] As soon as 1845, Charles Hancock (1800–1877) and Henry Bewley formed The Gutta Percha Company in the U.K. with their first product being a gutta percha stopper for soda water bottles. Gutta Percha's useful properties were its hardness, its electrical insulating behavior, its extremely low coefficient of thermal expansion, and its low softening temperature of around 70 °C. These properties meant that Gutta Percha was the first commercially utilized *natural* thermoplastic material, that is, it could be heated and cast continuously. It rapidly gained use in a large number of applications such as for golf balls, furniture, utensils, as imitation leather, in ornaments and jewelry, for various surgical devices, and as an

---

[1] http://www.ashmolean.org/ash/amulets/tradescant/tradescant02.html

[2] http://plastiquarian.com/?page_id=14213

© Springer International Publishing AG, part of Springer Nature 2018
I. Baker, *Fifty Materials That Make the World*,
https://doi.org/10.1007/978-3-319-78766-4_17

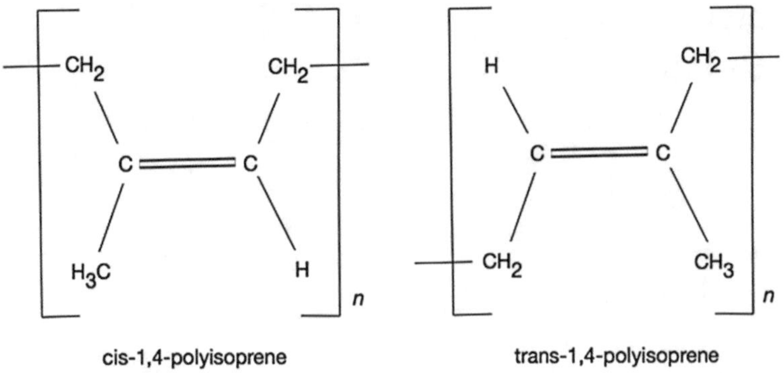

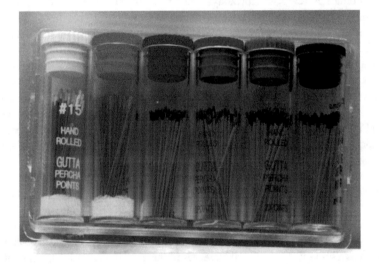

cis-1,4-polyisoprene   trans-1,4-polyisoprene

Fig. 17.1 Molecular structure of the polymers Gutta Percha (left) and latex (right). The two polymers are chemically similar but the chemical groups are arranged differently in the molecules

Fig. 17.2 Gutta percha "points" which are used to fill root canals in teeth. (Courtesy of Crystal Wild)

electrical insulator [2]. Although few people nowadays have heard of Gutta Percha, in the nineteenth century it was a household name.

As an electrical insulating material, Gutta Percha solved an immediate problem of the need for an insulator for underwater telegraph cables. It was first used in 1850 on the first submarine cable that was laid between England and France, a mere 32 km, by Briton John Watkins Brett's (1805–1863) of the Anglo-French Telegraph Company.[3] Only 8 years later in 1858, the first 3200 km transatlantic cable was laid between Ireland and Newfoundland by the American Telegraph Company, which

---

[3] http://www.gracesguide.co.uk/John_Watkins_Brett

was created by the American entrepreneur Cyrus West Field (1819–1892).[4] Sadly, the cable only worked for 3 weeks. British interests dominated the submarine cable industry until the twentieth century because of the needs for rapid communication in the global British Empire. Indeed, at the time, locations in the British Empire were the only source of Gutta Percha. Gutta Percha was the material of choice for this application up until the late 1930s when it was replaced by the newly-invented material polyethylene.[5]

In another unique application, Gutta Percha changed golf into a sport for the masses. Gutta Percha golf balls, or Gutties as they were known, were introduced in 1848. At only one shilling, they were much cheaper to make than earlier golf balls, which were initially made of wood and then later were made of leather stuffed with feathers.[6] The use of gutties continued until 1900, when they were replaced by balls made of natural rubber.

As noted at the beginning of this chapter, one of the few current uses for Gutta Percha is for filling root canals, but even that use is threatened by newer resins [3], see Fig. 17.2. Gutta Percha appears to be a material of the past.

References

1. Khatavkar, R. A., & Hegde, V. S. (2010). The weak link in Endodontics: Gutta-Percha-a need for change. *World Journal of Dentistry, 1*(3), 217–224.
2. Amato, I. (1998). *Stuff: The materials the world is made of.* New York: Avon Books, Inc. ISBN-10: 0380731533.
3. Prakash, R., Gopikrishna, V., & Kandaswamy, D. (2005). Gutta-Percha: An untold story. *Endodontology, 17*(2), 32–36.

---

[4] https://www.britannica.com/biography/Cyrus-W-Field

[5] The Plastics Historical Society.pdf

[6] http://www.scottishgolfhistory.org/origin-of-golf-terms/golf-ball-feathery-gutty-haskell/#Gutty

Chapter 18
Iron

Iron, a lustrous silvery, malleable, very ductile metal, exists in three different crystal structures or allotropes: up to 912 °C it is body centered cubic (b.c.c.); from 912-1394 °C it is face centered cubic (f.c.c.); and from 1394 °C up to the melting point of 1538 °C it is again b.c.c., see Fig. 18.1. Iron is the fourth most abundant element in the Earth's crust at 6.3% and the second most abundant metal after aluminum at 8.1%. In addition, the Earth's inner and outer cores, which constitute 35% of the mass of the planet, are composed of iron and nickel.

Iron is one of only three metals that are commonly found in their elemental state on Earth. Most of the elemental iron arises from meteorites in one of four different iron-nickel alloys with nickel contents from 5 to 57 weight percent. 100 tonnes of meteoroids, largely dust and gravel, hit the Earth's atmosphere every day,[1] some of which strike the Earth's surface whence they are called meteorites.[2] Perhaps the most interesting alloy identified in meteorites is the ferromagnetic tetragonal-structured Tetrataenite NiFe, see Fig. 18.2 [1–8], which is f.c.c. at high temperature. It is very difficult to transform the high-temperature f.c.c. structure to the room-temperature equilibrium Tetrataenite because the very low transformation temperature of ~330 °C means that the transformation takes a very long time. In meteorites, even though the temperature is likely quite low, Tetrataenite has the time for this transformation to occur. We know that the oldest iron objects, such as the beads found in Gerzah, Lower Egypt that date to 3500 B.C. [9], were made from meteoric iron as they are identifiable from their iron-nickel chemistry [10]. These objects were simply pieces of meteorite that were beaten into shape, possibly with intermediate heating to soften them, to allow continued working. The dagger found with the mummy of King Tutankhaman, who briefly reigned from 1332–1323 B.C. in Egypt, contains more than 10% nickel indicating that it was produced from meteoric iron [11].

---

[1] https://science.nasa.gov/science-news/science-at-nasa/2011/01mar_meteornetwork

[2] https://www.britannica.com/demystified/whats-the-difference-between-a-meteoroid-a-meteor-and-a-meteorite

© Springer International Publishing AG, part of Springer Nature 2018
I. Baker, *Fifty Materials That Make the World*,
https://doi.org/10.1007/978-3-319-78766-4_18

Fig. 18.1 The unit cells of (**a**) B.C.C. iron, and (**b**) F.C.C. iron

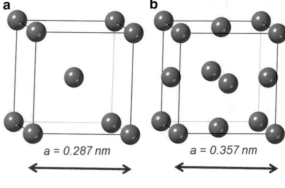

Fig. 18.2 The unit cell of tetragonal NiFe, which is called Tetrataenite

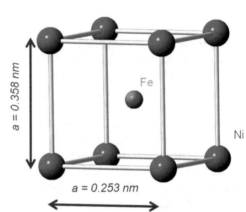

Naturally-occurring non-meteoric elemental iron, referred to as Telluric Iron, is found at a very few locations on Earth. The only significant deposit is at Disko Bay, Greenland.[3] This telluric iron exists as small iron grains containing up to 4 weight percent nickel in large basalt boulders that weigh from a few to tens of tonnes. The local Inuit, who were the only people to utilize telluric iron, obtained the iron grains by crushing the basalt and then hammering them into discs that were then inserted into bone handles to make knives.

The smelting of iron from its ore appears to have started in Anatolia by proto-Hittites [12]. One possible site is Kaman-Kalehöyük, in modern day turkey, where steel was found dating to 1800 B.C. [13]. Other possibilities are the states of Sumer, Akkadia and Assyria in Mesopotamia where ore smelting could have started as early as 3000 B.C. [14]. We will likely never know for sure where iron production from its ores such as the oxides hematite (Fe_2O_3) and magnetite (Fe_3O_4) started. But by 1200 B.C. smelted objects are found in the Middle East, South East Asia and South Asia.

---

[3] https://www.revolvy.com/topic/Telluric%20iron&item_type=topic

The start of the Iron Age, when iron production became widespread in significant quantities, is generally considered to be from around 1200–600 B.C., the exact start date depending on the region of the World. In Europe the start of the Iron Age corresponds to the Bronze Age Collapse of roughly 1200–1150 B.C. when most cities in the Eastern Mediterranean were destroyed due possibly to multiple man-made and natural disasters [15]. It has also been suggested that mysterious Sea Peoples who invaded may have cut off trade routes that supplied the tin that was needed to make Bronze.[4] The oldest iron artifacts from China date to around 500 B.C. [16]. In Africa the Iron Age succeeded the Stone Age, with no intervening Bronze Age.

Why was Bronze (a copper-tin alloy) produced before Iron, even though iron is more than a thousand times more common than copper and over 10,000 time more common than tin - and copper and tin are almost never found together? The answer is the temperature required to smelt the material. Copper melts at 1085 °C, a temperature reached by Ancient pottery kilns, whereas iron melts at 1538 °C, a difficult temperature to reach in the Ancient World. However, for additions of carbon up to 4.3 wt. % the melting temperature of iron-carbon alloys decreases to the much more attainable 1147 °C. The iron so produced was a similar strength to bronze although the density of Iron (7870 kg m^{-3}) can be lower than bronze (7400–8900 kg m^{-3}).

When iron oxide is heated in a furnace with charcoal, the charcoal provides heat and reduces the iron oxide, via a complex series of reactions, to iron, producing carbon dioxide in the process. When the furnace cools, the result is a spongy, porous mass of iron and slag. The iron is then hammered or wrought, which drives out the cinders and slag. The result, referred to as *wrought iron*, contains 0.02–0.08 wt. % carbon - the solubility of carbon in b.c.c. iron is very low, only 0.021% even at 900 °C. Wrought iron was the most common form of iron throughout the Iron Age. Modern wrought iron has a relatively low yield strength compared to steels of 159–221 MPa, but can show significant tensile elongation to failure. Steel, which contains 0.3–1.2 wt. % carbon, was first produced in Western Persia around 1000 B.C., but did not become common until the 1850s (see chapter on steel). During the Iron Age, the best weapons were made from steel, which has a much higher strength than wrought iron. The problem was that steel was hard to make because there was no way to control the carbon content. Wrought iron was, thus, easier to make, and, hence, much cheaper and much more common [17].

Initially, cupola furnaces were used to smelt iron, but by the first century A.D., blast furnaces, in which air is forced through the furnace from the bottom using bellows, were being used in China. By the late middle ages, 3 m high, chimney-like blast furnaces were being built. The molten metal ran out from the bottom of the furnace into containers called "pigs", the unusual name "pig" arising because the containers awaiting their iron diet were reminiscent of suckling pigs. At very high temperatures, the iron dissolved 3–4.5% carbon, which lowers the melting point – the lowest melting point occurs for iron with 4.3 wt. %, which melts at 1147 °C. The resulting "pigs" contain impurities such as sulfur, phosphorus and manganese. Removing impurities results in *Cast Iron*, which can be used for stoves, lampposts,

---

[4] http://study.com/academy/lesson/iron-vs-bronze-history-of-metallurgy.html

rails, and radiators. This is quite different to wrought iron because the carbon makes the cast iron quite hard but very brittle. The production of cast iron differs little today although the furnaces used are much bigger and their control is much more precise.

A number of technological innovations occurred in the U.K. during the Industrial Revolution that improved the efficiency and production of iron. In 1709, Abraham Darby (1678–1717) demonstrated that coke, a hard, porous, low-impurity, high-carbon fuel produced from bituminous coal baked to remove impurities, could be used instead of charcoal, which was generated from wood. This may not seem like a big deal, but the U.K.'s trees were rapidly dwindling and coke provided a reliable source of carbon. The result was cheaper cast iron that was a huge impetus to the Industrial Revolution. Coke-produced cast iron was eventually used for rails, steam engines, ships, aqueducts, cannons, cannonballs, and numerous other uses. The mass-produced iron cannons that were developed in England were not better than the existing bronze cannons and they were heavier, but they were much cheaper to mass-produce. Thus, this contributed to the dominance of the Royal Navy. Abraham Darby's grandson Abraham Darby III (1750–1789) built the first bridge made of iron in 1779. The 30-metre bridge spans a river gorge at what is now called Ironbridge in Shropshire, England. Since no one had built a bridge made out of iron before, the joints were basically those used in carpentry. In 1783, Englishman Henry Cort (1741–1800) developed the Puddling Furnace, in which the molten metal is stirred by strong workers enabling uniform oxidation of the furnace charge. The carbon was thus removed. As the carbon content decreases the melting point of the iron increases and some of the iron solidifies and is removed from the furnace. The resulting iron could be forged and rolled. Towards the end of the eighteenth century, cast iron replaced wrought iron since it was cheaper. Again, providing impetus to the Industrial Revolution. The expansion of railways from the 1830s onwards greatly increased the use of iron. Iron was first used for ship's fittings on the Royal James, a British warship launched in 1670, and from the early eighteenth century French and later British Warships used increasing amounts of iron in fittings. At least in the U.K. this was partly driven by a shortage of timber. This eventually led to the first iron barge that was built by John Wilkinson (1728–1808) in 1787 and launched on the River Severn, Shropshire, England, while the first iron-hulled war-ship was the *Nemesis*, built for the British East India Company in 1839. In 1843, the S.S. Great Britain, designed by Isambard Kingdom Brunel (1809–1859), the world's first ocean-going ship that combined steam-power, a screw propeller and an iron hull was launched. She was the longest passenger ship at the time, an achievement she retained until 1854.

B.C.C. iron is one of three elements that are ferromagnetic at room temperature: the others are hexagonal close packed (h.c.p.) cobalt and face centered cubic nickel. (Gadolinium, terbium, dysprosium and holmium are also ferromagnetic but only at temperatures below room temperature of 20 °C, -54 °C, -188 °C, and -254 °C, respectively.) These have Curie Temperatures, the temperature at which thermal fluctuations prevent alignment of the magnetic moments so that the ferromagnetic behavior is lost, of 770 °C, 1131 °C and 358 °C, respectively. Iron is the strongest elemental magnet with a saturation magnetization of 2.15 Tesla compared to 1.79 Tesla and 0.61 Tesla for cobalt and nickel, respectively. All three are all soft magnets, that is, they are easily demagnetized.

Iron can display a range of strengths depending on its purity and form. High-purity bulk single crystals of iron have a very low yield strength of 10 MPa. However, defect-free single crystal whiskers of iron have been shown to have strengths of 11,000 MPa, which is close to the predicted theoretical strength [18]. Fine-grained iron can have strengths of 340 MPa, whereas cold working of polycrystalline iron can produce strengths of 690 MPa.

There are several forms of modern cast iron. The two primary forms are gray cast iron and white cast iron. In gray cast iron, the carbon is present in the form of graphite flakes, which are in thermodynamic equilibrium. These flakes provide a self-lubricating behavior and so gray cast iron is very good in applications where wear is an issue, for example in engine blocks, gearbox cases, flywheels, and machine-tool bases. The size and shape of the graphite particles control the mechanical properties. Gray cast iron is strong in compression but has very limited ductility in tension (0.5% elongation) showing a tensile strength of 345 MPa. Gray cast irons, which consist primary of iron, carbon, and manganese and up to 3 wt. % silicon, are the most common cast iron.

In white cast iron, the carbon is present as the metastable iron carbide (Fe_3C), called cementite. These cast irons have low silicon since the silicon forces the carbon out of solution producing the equilibrium graphite phase instead of cementite. White cast irons must also be cooled more rapidly than gray cast irons to avoid the equilibrium graphite forming, which limits the size of parts that can be made of white cast iron. Adding chromium allows much larger castings to be made, up to several tonnes, and the chromium leads to the formation of chromium carbides, which increases the already good abrasion resistance further. White cast irons are quite brittle, showing no elongation before tensile failure and have tensile strengths only half that of gray cast irons of 175 MPa. These do not find many applications but are used on some bearing surfaces.

The reasons for the name "white" and "gray" is that these are the colors of the fracture surface, the white appearance is due to the cementite. Malleable cast is white cast iron that has been heat treated so that the cementite turns into spheroids of the equilibrium graphite phase. Malleable cast irons have similar tensile strengths to gray cast iron but can show elongations up to 12%, and have applications as axle bearings and automotive crankshafts. Malleable cast iron dates back to at least the fourth century B.C. China [19]. The last major form, ductile or nodular cast iron is a newer form of cast iron that was invented in 1943 by the American Keith Dwight Millis (1915–1992)[5] and co-workers at the Canadian International Nickel Company (INCO). He added small amounts (0.02–0.1%) of magnesium (cerium is now added as well), which caused the graphite to form nodules [20], see Fig. 18.3. Ductile cast iron not only shows a higher tensile strength of around 490 MPa but also exhibits

---

[5] https://rpi.edu/about/alumni/inductees/millis.html

Fig. 18.3 Optical
Micrograph of ductile cast
iron showing dark graphite
nodules in a b.c.c. ferrite
matrix. (Courtesy of
H.N. Han)

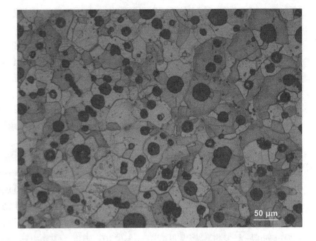

Fig. 18.4 Frying pans and
other kitchenware are often
made of cast iron

around 18% elongation. Ductile cast irons are used for gears, crankshafts and cam-
shafts. Cast iron is also used to make objects such as frying pans, see Fig. 18.4.

A problem with iron is that it oxidizes to give reddish-brown "rust" of Fe_2O_3.
xH_2O. A well-known exception to this is the 7 m high Iron Pillar of Delhi, which has
stood since the end of the fourth century A.D. This does not rust for a couple of
reasons. First, because of the high levels of sulfur in the iron, which produces a rust
resistant layer on the surface [21] - unfortunately, this is not a good solution to iron's
rusting problem since sulfur is very deleterious to the mechanical properties of iron.
Second, the local climate is such that heating of the pillar during the day prevents
condensation onto the pillar at night [22].

Currently, 90% of all the metal that is refined is iron. In the developed world,
there is almost nothing cheaper that you can buy than iron or steel at around $0.5 per
kilogram. The abundance and low cost mean that there is 2200 kg of iron per person
in use today: in the developed world that amount jumps to 7000–14,000 kg of iron

per capita. In 2015, the World production of "pig" iron was 1180 million tonnes, of which China produced a stunning 710 million tonnes (60%). The U.K., the place where the Industrial Revolution and mass iron production started, produced only 9 million tonnes (0.8% of the World output). China's production may have peaked, but other large developing countries such as India and Brazil are likely to increase their production in future.

References

1. Albertsen, J. F., Jensen, G. B., & Knudsen, J. M. (1978). Structure of Taenite in two Iron meteorites. *Nature, 273*(8), 453–454.
2. Albertsen, J. F., Aydin, M., & Knudsen, J. M. (1977). Mossbauer effect studies of Taenite lamellae of an Iron meteorite Cape York (III.A). *Physica Scripta, 17*, 467–472.
3. Nagata, T. (1983). High magnetic Coercivity of meteorites containing the ordered FeNi (Tetrataenite) as the major ferromagnetic constituent. *Journal of Geophysical Research, 88*, A779–A784.
4. Rubin, A. E. (1994). Euhedral Tetrataenite in Jelica meteorite. *Mineralogical Magazine, 58*, 215–221.
5. Petersen, J. F., Aydin, M., & Knudsen, J. M. (1977). Mossbauer spectroscopy of an ordered phase (superstructure of FeNi in an Iron meteorite). *Physics Letters, 62A*(3), 192–194.
6. Scott, E. R. D., & Rajan, R. S. (1981). Metallic minerals, thermal histories and parent bodies of some Xenolithic, ordinary Chondrite meteorites. *Geochimica et Cosmochimica Acta, 45*, 53–67.
7. Clarke, R. S. (1980). Tetrataenite - Ordered FeNi, a new mineral in meteorites. *American Mineralogist, 65*, 624–630.
8. Scott, E. R. D. (1979). Identification of clear Taenite in meteorites as ordered FeNi. *Nature, 281*, 360–362.
9. Rehren, T., Belgya, T., Jambon, A., Káli, G., Kasztovszky, Z., Kis, Z., Kovács, I., Maróti, B., Martinón-Torres, M., Miniaci, G., Pigott, V. C., Radivojević, M., Rosta, L., Szentmiklósi, L., & Szőkefalvi-Nagy, Z. (2013). 5,000 years old Egyptian iron beads made from hammered meteoritic iron. *Journal of Archeological Science, 40*, 4785–4792. https://doi.org/10.1016/j.jas.2013.06.002.
10. Wayman, M. L. (1989). Neutron activation analysis of metals: A case study. In S. J. Fleming, H. R. Schenck (Eds.), *History of technology: The role of metals edited by*. The Museum Applied Science Center for Archeology, The University Museum, University of Pennsylvania. ISSN 0198-0106.
11. *EOS, Earth and Space Science News*, 15 July 2016, p. 4
12. Spoerl, J. S. (2004). A brief history of iron and steel production. Manchester, NH: Saint Anselm College, Online Guide.
13. Akanuma, H. (2005). The significance of the composition of excavated iron fragments taken from stratum III at the site of Kaman-Kalehöyük, Turkey. *Anatolian Archaeological Studies., 14*, 147–158.
14. Weeks, M. E. (1932). Discovery of the elements. I. Elements known to the ancient world. *Journal of Chemical Education, 9*, 4.
15. Cline, E. H. (2014). *1177 B.C.: The year civilization collapsed*. Princeton: Princeton University Press. ISBN 9780691140896.
16. Wagner, D. B. (1993). *Iron and steel in ancient China*. Leiden: E. J. Brill. ISBN9009096329.
17. Amato, I. (1998). *Stuff: The materials the world is made of*. New York: Avon Books, Inc. ISBN-10: 0380731533.

18. Brenner, S. S. (1956). Tensile strength of whiskers. *Journal of Applied Physics, 27*, 184–1491.
19. Wagner, D. B. (1993). *Iron and steel in ancient China*. Leiden: E. J. Brill. ISBN 978-90-04-09632-5.
20. Millis, K. D., Gagnedin, P., & Pilling, N. B. (1949). *Cast Ferrous Alloy*. US Patent 2485760, issued October 25.
21. Chaline, E. (2012). *Fifty minerals that changed the course of history*. Buffalo: Iron Firefly Books Ltd..
22. Elmsley, J. (2001). *Nature's building blocks. An A-Z guide to the elements*. Oxford: Oxford University Press.

Chapter 19
Kevlar and Other Aramid Fibers

The name poly-paraphenylene terephthalamide, see Fig. 19.1, doesn't exactly roll of the tongue, which is presumably why Dupont gave it the much cooler commercial name of Kevlar. Kevlar was invented by Stephanie Louise Kwolek (1923–2014)[1] and Paul Wintrop Morgan (1911–1992)[2] at E. I. du Pont de Nemours and Company for which US patent 3287323A "Process for the production of a highly orientable, crystallizable, filament forming polyamide" was awarded in 1966 [1, 2]. It is produced by the condensation reaction between 1,4-phenylene-diamine and terephthaloyl chloride: the condensate, that is the chemical left over from the reaction, is hydrochloric acid. Mechanical drawing of the product produces Kevlar fibers. Kevlar can be considered one of a family of **ar**omatic poly**amid**e, which is usually contracted to "aramid", fibers that include Nomex [3], Technora,[3] Teijinconex and Twaron.[4] The invention of Kevlar followed earlier work by these two American chemists who filed a patent with Richard Sorenson Wayne (1926-) awarded in 1962 for "Process of making wholly aromatic Polyamides". Kevlar fibers are typically spun into tows or sheets for use in composites typically with an epoxy matrix.

The Federal Trade Commission's has a specific definition for an aramid fiber: "A manufactured fiber in which the fiber-forming substance is a long-chain synthetic polyamide in which at least 85% of the amide linkages, $(-CO-NH-)$, are attached directly to two aromatic rings". The amide group is also the basis of the polyamide polymers called nylons, which do not contain aromatic rings.

Aramid fibers as a whole have a number of excellent properties including being non-conductive, having low flammability, good abrasion resistance and good resistance to organic solvents. Nomex, also developed by DuPont in the early 1960's, was the first of the aramids and was marketed in 1967. Nomex is chemically identical to Kevlar but the backbone chain is arranged differently so that it is not

---

[1] https://www.chemheritage.org/historical-profile/stephanie-l-kwolek

[2] https://www.nap.edu/read/4779/chapter/31

[3] http://www.teijinaramid.com/brands/technora/

[4] http://www.teijinaramid.com/brands/twaron/

© Springer International Publishing AG, part of Springer Nature 2018
I. Baker, *Fifty Materials That Make the World*,
https://doi.org/10.1007/978-3-319-78766-4_19

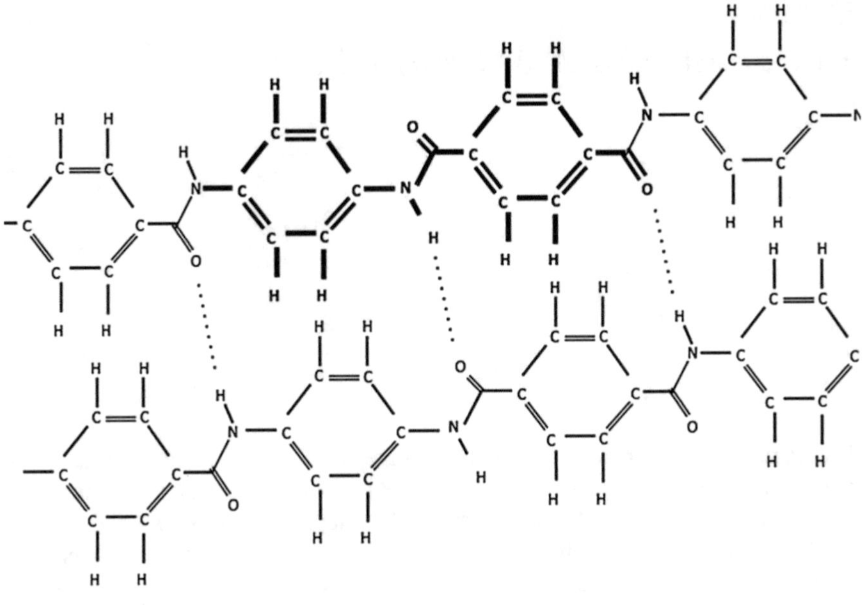

Fig. 19.1 The structure of poly-paraphenylene terephthalamide or Kevlar. The black section highlights the repeat unit that contains two aromatic carbon rings in the polymer. The individual polymer chains are held together by hydrogen bonding indicated by the dotted lines

crystalline and much weaker than Kevlar.[5] However, Nomex has excellent resistance to heat, chemicals and radiation. Thus, it is used in numerous applications for firefighting and in fire-resistant clothing for users such as racing car drivers and military personnel. Its fire resistance and ability to withstand extreme environments has seen its use in various space exploration applications, and its good acoustic properties are utilized in loudspeaker drivers.

As outlined in the original patent, the polymer chains are highly aligned in Kevlar and the fibers are crystalline; Twaron developed only a little later than Kevlar by the Dutch company AKZO has a very similar structure. This gives Kevlar and Twaron fibers excellent mechanical properties. For example, Kevlar 149, which has a density of 1470 kg/m^3, has a tensile elastic modulus (resistance to stretching) of 186 GPA, a tensile strength of 3400 MPa but an elongation to failure of only 2%. While the very best ultra high strength steel would exhibit a similar elastic modulus and elongation to the Kevlar, the tensile strength would only be around 1700 MPa (most steels have much lower strength). However, since the density of steels is around 8000 kg/m^3, on a density-compensated basis, the specific tensile strength of the ultra high strength steel is 210 kN · m/kg, while that of the Kevlar 149 is 1950 kN · m/kg, or almost 10 times higher.

Of course, Kevlar fibers are not used alone but are incorporated into a polymer matrix that protects the fibers, see Fig. 19.2. Thus, the strength is a combination of

---

[5] http://www.pslc.ws/macrog/aramid.htm

Fig. 19.2 A swatch of
Kevlar

that of the fibers and that of the matrix, which is typically epoxy. Epoxy has a much
lower modulus (2.4 GPa) and tensile strength (27–90 MPa) than Kevlar fibers. Thus,
a typical Kevlar-epoxy composite with 60% Kevlar fibers aligned in one direction
has a modulus of 76 GPa in the fiber direction and 5.5 GPa transverse to the fibers
and a tensile strength of 1380 MPa in the fiber direction and 30 MPa transverse to
the fibers [4]. However, the density of epoxies is little lower than that of Kevlar and,
thus, the density of a composite containing 60% Kevlar fibers is about $1400\,kg \cdot m^{-3}$.

The original use envisioned for Kevlar was to replace steel in tires, for which it
is used extensively as well as in the brakes and bodies of cars. Since it's introduction
it has garnered a large number of applications in boats and planes where its high
strength, toughness and low density are critical. Kevlar has also become the preva-
lent fiber for high-performance racing sails not only because of its high strength but
also because of its resistance to degradation by ultraviolet light. Probably the best
known usage is in bulletproof vests and helmets, and in knife-proof body armor,
where its energy absorption is the key property. Kevlar composites are also used in
various sports equipment including snowboards, tennis strings and hockey sticks.
No doubt that aramid composites will continue to find applications that utilize their
high strength, high toughness and high modulus coupled with low density and thermal dimensional stability.

References

1. van Dulken, S. (2000). *Inventing the 20th century: 100 inventions that shaped the world. From the airplane to the zipper.* New York: New York University Press. ISBN: 0–8147–8808-4.
2. van Dulken, S. (2000). *Inventing the 20th century: 100 inventions that shaped the world from the airplane to the zipper.* New York: New York University Press. ISBN-13: 978-0814788127.
3. Kwolek, S., Mera, H., & Takata, T. (2002). High-performance fibers. In *Ullmann's encyclopedia of industrial chemistry.* Weinheim: Wiley-VCH. https://doi.org/10.1002/14356007.a13_001.
4. Callister, W. D., Jr. (2001). *Fundamentals of materials science and engineering.* Hoboken, NJ: Wiley, Inc. ISBN0-471-39551-X.

Chapter 20
Lead

Lead is a bluish-white face centered cubic (f.c.c.) metal that usually looks gray due to the rapid formation of an oxide or a carbonate on the surface, see Fig. 20.1. Lead appears to be the first metal smelted from its ore, typically galena. The earliest known site, which dates from the early sixth millennium B.C., is Çatalhöyü in current-day Turkey [1]. Both sculptures and coins made from lead have been found in Egyptian tombs dated to 5000 B.C. [2]. The Romans used lead extensively for utensils, drinking vessels and bowls, ballistics, cosmetics and for plumbing. The word plumbing is derived from the Latin for lead *Plumbum*, which is also the origin of the chemical symbol for lead of Pb and of *plumb bob* and *plumb line* (a weight attached to a line), which are used for determining verticals or the depth of water. Many of these uses meant that Romans were sure to have ingested significant amounts of lead. Worst was probably that they used lead pots to boil down grape juice, which was used as a sweetener and preservative in wine and food.[1] Perhaps, not surprisingly, analysis of lead in skeletons from Imperial Rome showed most had lead levels above 1 part per million (p.p.m.), the World Health Organization (WHO) level of concern, while some exceeded 10 p.p.m., which constitutes severe lead poisoning according to WHO [3]. The ingestion of lead was probably worse amongst richer Roman citizens who could drink more wine and afford lead cups and vessels.

The extensive use of lead arose for a number of reasons. First, it is readily extracted from its shiny metallic-looking ore, galena (lead sulfide) by grinding, using flotation processing, oxidation to remove impurities such as sulfur, and finally by smelting with charcoal. Since Roman times lead has also been an important byproduct from the refining of silver[1] from which it was mostly obtained in that period. Second, lead is quite soft and malleable, and, hence, easy to form. This is because at room temperature lead is at half of its absolute melting temperature of 600.6 Kelvin or 327.5 °C – the properties of materials depend on the homologous temperature, that is on the fraction of the melting temperature on the absolute

---

[1] http://ancientstandard.com/2010/01/01/would-you-like-some-lead-with-your-food/

Fig. 20.1 The face centered cubic crystal structure of lead

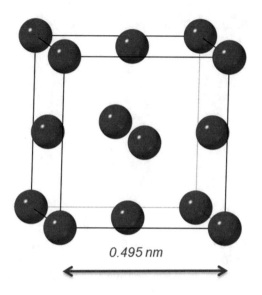

0.495 nm

temperature scale, which starts at −273.15 °C or 0 Kelvin. Third, lead is quite resistant to corrosion, and, thus, is often used in chemical reactors or for lining such vessels.

Lead accumulates in the body, producing lead poisoning, whose symptoms include developmental delays in children, abdominal pain, fatigue, insomnia, irritability, brain damage, kidney damage, neuropathy and neuromuscular problems. At very high levels, it can lead to seizures and death: 853,000 deaths were attributed to lead in 2013.[2,3] Organic forms of lead are more toxic than inorganic forms. At least three of the Franklin expedition that set off from the U.K. to find the North-West passage in 1845 died of lead poisoning from the lead-tin solder that was used to seal the tin food cans with which they were provisioned [4].

Unfortunately, the toxicity of lead wasn't recognized until the second half of the twentieth century, but lead poisoning has been a problem since lead was first used. A problem that continues to this day. Starting with the Romans, lead has been used extensively for plumbing, and while lead use in plumbing was largely curtailed after World War II, lead-tin solders were used to join the replacement copper tubing (copper has been largely replaced in new plumbing by polyvinyl chloride due to the rise in the price of copper). Lead was used as a pigment in paint from the fourth century B.C.[4] It was used both for painting buildings, furniture and even children's toys. Lead poisoning is particularly an issue for small children who may ingest chips or flakes of paint. The use of lead paint has now largely been curtailed apart from paints used by some artists.

[2] https://www.lead.org.au/history_of_lead_poisoning_in_the_world.htm

[3] http://www.who.int/mediacentre/factsheets/fs379/en/

[4] http://www.webexhibits.org/pigments/indiv/history/leadwhite.html

Although these are age-old problems with lead, modern society produced its own lead issues. In the early 1920s, while working at General Motors' Dayton Research Laboratories, Thomas Midgley, Jr. (1889–1944) developed tetraethyl lead (TEL), $(Ch_3Ch_2)_4Pb$, which was used as an anti-knock agent in internal combustion engines (ICE)[5] [5, 6] even though ethanol could be used for the same purpose. TEL both raises the octane of gasoline or petrol allowing higher compression engines and increased fuel efficiency, and prevents valve wear in the ICE. Adding TEL to gasoline may be the most efficient way of contaminating the environment with lead – and consumers paid to do it. TEL started to be phased out in the late 1980s and is now banned in most countries in the world[6] because of its toxicity. It also poisons the catalytic convertors on cars.

Other environmental issues with lead include the use of lead in shot for fishing and hunting. A huge leap forward in mass producing uniform lead shot, a drop tower, was patented by Bristol plumber William Watts of Bristol, England in 1782 [7] - a technique that has subsequently been used, with tweaks, for other metals and still is. In this process, molten lead is dropped a large distance into water. During the fall the lead droplets form spheres in order to minimize their surface area and solidify. The water simply cools the droplet. The original incarnation of the drop tower was built at Watts' house where he ingeniously added a 9 m tower to the top of his house and a 10.5 m pit below his cellar to give a 27 m drop.[7] Can you imagine the permits and reviews that you would need if you wanted to do this today – "you want to do what to your house, Mr. Watts? You want to melt lead and do what?" Remarkably, the tower was used up until 1968 by the Sheldon Bush & Patent Shot Company. Lead shot for hunting waterfowl was banned in the U.S.A. in 1991 since the ingestion of the shot is extremely harmful, but it remains available for other uses. California banned the use of lead shot for all hunting which will be phased out by July, 2019 – steel shot can be used instead.

The major source of lead in landfills is electronic components and TV screens. Lead-tin solder, used to join electronic components, was banned in the European Union in 2006 and several lead-free solders have been developed [8]. Lead was added to the glass screens of cathode ray tube-based televisions (about 0.25 kg per television) in order to shield the viewer from the small amount of X-radiation produced. Lead glass continues to be used for radiation shielding, see Fig. 20.2.

It may sound like the use of lead has been an unmitigated disaster, but some uses have been vital. Johannes Gutenberg (1398 – February 3, 1468), the blacksmith, goldsmith, printer and publisher, who introduced printing to Europe, relied on an alloy consisting of lead, tin and antimony for his movable type.[8] The lead-acid battery, which was invented by French physicist Gaston Planté (1834–1889) in 1859, is the oldest type of rechargeable battery.[9] It accounts for 88% of the lead used in the

---

[5] https://www.britannica.com/biography/Thomas-Midgley-Jr

[6] https://www.britannica.com/science/tetraethyl-lead

[7] http://www.inventricity.com/local-heroes-william-watts

[8] https://www.britannica.com/biography/Johannes-Gutenberg

[9] https://www.britannica.com/biography/Gaston-Plante

Fig. 20.2 A lead glass window on an X-ray set is used so that the set operation can be observed while providing shielding from the radiation

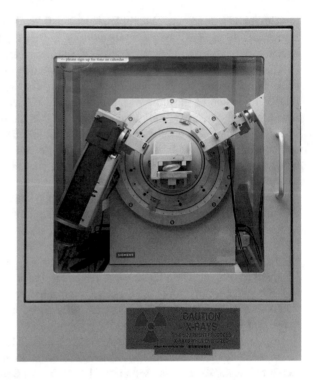

U.S.A. since they are the batteries used for starting cars. The battery consists of a lead anode (positive terminal) and a cathode (negative terminal) that consists of lead oxide packed into a lead grid. The low strength and low melting point (327 °C) of lead mean that in hot climates failure of the battery can occur by creep (continuous slow deformation that occurs under load at higher temperatures) of the lead electrodes under their own weight. Nearly 99% of lead used in batteries is recycled in Europe and the U.S.A. The weight of lead batteries is a disadvantage in most vehicle applications, but in forklift trucks the batteries are used as a counterweight to the object being lifted. Similarly, lead's high density combined with its corrosion resistance is the reason that it is used for ballast in sailboats.

Because of its corrosion resistance, lead sheets are used for roofing or flashing, particularly in Europe, for protective sheathing in underwater cables, and in chemical reactors. Because of its density and high number of electrons, lead is good at absorption of radiation and, thus, is often used for shielding for example at the dentist if you have an X-ray.

Other significant uses of lead are in some glass and ceramics, as an alloying component in other metals, and in ammunition. Lead is not, of course, used for the "lead" in a pencil, which is made of graphite. Also, one can't chemically transmute lead into gold and fulfill the Alchemists dreams.

The low room temperature yield strength of pure lead can be improved by adding ordered f.c.c. Pb_3Ca precipitates to the f.c.c. lead matrix produce an analog of a nickel-based superalloy as I demonstrated in the early 1990s. The yield strength can be increased from around 14 MPa for pure lead to 58 MPa for an alloy containing 70 volume percent Pb_3Ca, but with a concomitant reduction in ductility [9]. Oxide dispersion strengthening can also significantly improve the strength of lead.

The Roman production of lead of 80,000 tonnes per year was not reached again until the start of the Industrial Revolution. The first known lead mines date to 2000 B.C. in the Iberian Peninsula. More recently, lead production has grown from 3.4 million tonnes per year in 1990 to about 4.8 million tones in 2016[10] with China being by far the largest producer.[11] This only accounts for half of the lead used, since the rest is obtained from recycling: lead is one of the most recycled materials. This annual growth in production of around 1.1% is much below global GDP of ~3–3.8% per year. It's not clear that the use of lead will continue to grow at all. The growing use of electric cars and trucks, which currently use lithium ion batteries and may use more exotic batteries such as sodium ion batteries in future, may be the death knell for lead-acid batteries except for specialized applications. This may not be a bad thing because in addition to its toxicity, if lead is continued to be used at the current rate, lead will run out in 18–42 years [1].

References

1. Muhly, J. D. (1988). The beginnings of metallurgy in the old world. In R. Maddin (Ed.), *The beginnings of the use of metals and alloys, in the beginning of the use of metals and alloys.* Boston: MIT Press. ISBN: 13 9780262132329.
2. Stwertka, A. (2012). *A guide to the elements.* New York: Oxford University Press. ISBN: 978-0-19-983252-1.
3. Montgomery, J., Evans, J., Chenery, S., Pashley, V., & Killgrove, K. (2010). 'Gleaming, white, and deadly': Using lead to track human exposure and geographic origins in the roman period in Britain. *Roman Diasporas, Journal of Roman Archaeology, Suppl, 78*, 199–226.
4. Emsley, J. (2001). *Lead, nature's building blocks: An A-Z guide to the elements.* Oxford: Oxford University Press. ISBN: 0-19-850340-7.
5. Chaline, E. (2012). *Fifty minerals that changed the course of history.* Buffalo, NY: Firefly Book Ltd. ISBN 978-1-55407-984-1.
6. Markowitz, G., & Rossner, D. (2003). *Deceit and denial. The deadly politics of industrial pollution.* Berkeley: University of California Press. ISBN: 0-520-24063-4.
7. Guruswamy, S. (1999). *Engineering properties and applications of lead alloys.* Boca Raton: CRC Press.
8. Ganesan, S., & Pecht, M. G. (Eds.). (2006). *Lead-free electronics.* Hoboken: Wiley, Inc. ISBN: 978-0-471-78617-7.
9. Baker, I. (1993). Room temperature deformation of lead - Based "Superalloys". *Acta Metallurgica et Materialia, 41*, 2633–2638.

---

[10] https://minerals.usgs.gov/minerals/pubs/historical-statistics/global/

[11] https://minerals.usgs.gov/minerals/pubs/commodity/lead/

Chapter 21
Lead Zirconate Titanate

Ceramics are inorganic, non-metallic, solids that consist of a metal, a non-metal, a metalloid such as silicon, or a combination of these. Ceramics utilize ionic or covalent bonding, or both. When thinking of a ceramic you may think of pottery. This is perfectly reasonable since the word "ceramic" is derived from the Greek word *keramikos*, which means "of pottery" or "for pottery". "Keramic" is an old, rare, now disused spelling of "ceramic". The main commercial piezoelectric ceramic is Lead Zirconium Titanate with a typical composition of $PbZr_{0.52}Ti_{0.48}O_3$. It is referred to a PZT and adopts the perovskite crystal structure.

A few ceramics such as $MoSi_2$ and SiC are good electrical conductors and are used in applications such as heating elements in furnaces. Electrically-conductive ceramics, typically spinel crystal structures, based on various oxides of iron, cobalt, and manganese are used in thermally-sensitive resistors, or "Thermistors". These have electrical resistivities that vary strongly with temperature. A related application is in various sensors for gases such as oxygen (zirconia) or carbon monoxide (tin oxide). Indium tin oxide (ITO) is a transparent electrically-conductive oxide that has seen large increases in use as an optoelectronic material, and is widely used for flat panel displays.

By far the majority of ceramics are electrical insulators with a typical electrical resistivity of 1×10^{13} Ohms-m [1] which is 21 orders of magnitude lower that of an aluminum alloy at 4×10^{-8} Ohms-m. They are used in applications such as insulators in a spark plug or on an electrical power line, see Fig. 21.1. Dielectric ceramics (barium titanate, zirconium barium titanate, strontium titanate, calcium titanate and magnesium titanate) are also insulators that are used to make capacitors such as the multi-layer ceramic capacitors (MLCC) used in the electronics industry. In 2012, over a trillion MLCC were produced [2]. These compounds are used because of their high dielectric constant, which is a measure of how much the electrical permittivity (a measure of how a material is polarized by an applied electric field) can be increased compared to that of a vacuum.

Some dielectric ceramic crystals exhibit piezoelectricity - the word is derived from the Greek word *piezein*, meaning to press or squeeze. In 1880, the brothers

© Springer International Publishing AG, part of Springer Nature 2018
I. Baker, *Fifty Materials That Make the World*,
https://doi.org/10.1007/978-3-319-78766-4_21

Fig. 21.1 A spark plug. The white part is an insulating ceramic

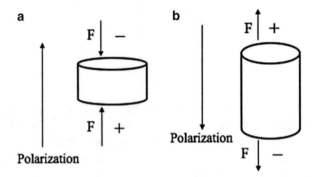

Fig. 21.2 In the Direct Piezoelectric Effect when a force, **F**, is applied to a piezoelectric crystal as in (**a**) positive and negative charges are produced on opposite faces of the crystal, leading to electrical polarization. In (**b**) a force, **F**, is applied in the opposite sense, producing the opposite charges on the faces and the opposite polarization

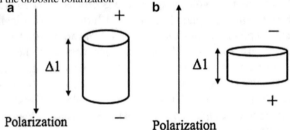

Fig. 21.3 In the Converse Piezoelectric Effect an electrical field applied to a crystal, as shown in (**a**), produces an increase in length (Δl), whereas an electric field applied in the opposite direct, as shown in (**b**), produces a contraction (Δl)

Jaques (1856–1941) and Pierre Curie (1859–1906) discovered the Direct Piezoelectric Effect in a variety of material including crystals of tourmaline, quartz, topaz, Rochelle salt and even cane sugar. Piezoelectricity occurs in many natural materials including bone and some proteins. The Direct Piezoelectric Effect is illustrated in Fig. 21.2. When a force is applied to a piezoelectric crystal as shown in Fig. 21.2a positive and negative charges are produced on opposite faces of the crystal, leading to electrical polarization. If a force is applied in the opposite sense, as shown in Fig. 21.2b, opposite charges appear on the faces and the opposite polarization results. In 1881, the Converse Piezoelectric Effect, illustrated in Fig. 21.3 was predicted by the Nobel-prize-winning physicist Gabriel Lippmann (1845–1921),

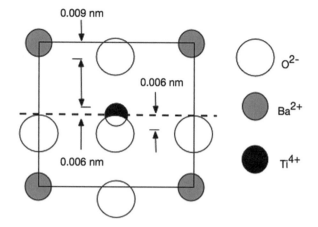

Fig. 21.4 In the tetragonal form (the vertical crystal axis is longer that the two other crystal axes) of barium titanate, which is stable below 120 °C, the centers of the atoms are slightly offset, which leads to an electrical dipole when a force is applied

and confirmed shortly thereafter by the Curie brothers. In the Converse Piezoelectric Effect, an electrical field applied to a crystal, as shown in Fig. 21.3a, produces an increase in length, whereas an electric field applied in the opposite direct produces decrease in length as shown in Fig. 21.3b.

Not all dielectric ceramics exhibit piezoelectricity, only those that do not have a center of symmetry can be piezoelectric. The first and archetypal piezoelectric ceramic is tetragonal barium titanate, see Fig. 21.4. In this crystal the centers of the atoms are slightly offset. Thus, when a stress is applied a charge develops in the crystal.

For many years, piezoelectricity was simply a curiosity. In 1917, the French physicist Paul Langevin (1872–1946) developed the first application which was an ultrasonic submarine detector that used quartz crystals. Nowadays, piezoelectric ceramics are used as: sensors (pressure sensors, strain sensors, guitar pickups); actuators (loudspeakers, inkjet printer heads, piezoelectric fuel injectors for diesel engines); transducers, which are both sensors and actuators (medical ultrasound, sonar systems, nondestructive testing), piezoelectric motors (autofocus in cameras), fixed frequency generators (quartz clocks, frequency standards in radio transmitters and receivers, generating clock pulses in computers), and voltage generators (cigarette lighter). The main piezoelectric ceramic that is used commercially is lead zirconate titanate with a typical composition of $PbZr_{0.52}Ti_{0.48}O_3$, which has the so-called perovskite crystal structure, see Fig. 21.5. It is commonly referred to as PZT, which is derived from the chemical symbols of the elements involved Pb, Zr and Ti. The first stable formulations of PZT were developed in the 1950s and 1960s, and mass production started in the late 1950s. Barium titanate, strontium titanate and quartz are also used as piezoelectrics.

Lead zirconate titanate is also a ferroelectric material. Ferroelectric crystals, which must again not have a center of symmetry, display a spontaneous charge, which can be reversed by the application of an electric field. Their usefulness lies in that they show a non-linear relationship between the applied electric field and the electric polarization of the crystal, see Fig. 21.6. Ferroelectric ceramics are used in

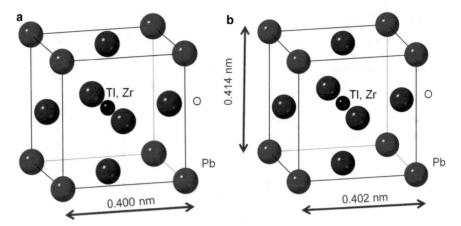

Fig. 21.5 Lead zirconate titanate (PZT), PbZr$_{1-x}$TixO$_3$, crystal exists as (**a**) a cubic crystal structure at high temperature and (**b**) as a tetragonal crystal structure at low temperature [3, 4]. The two
crystal structures are very similar, but in the cubic structure all axes in the crystal have the same
length, whereas in the tetragonal crystal structure one axis is of different length to the other two

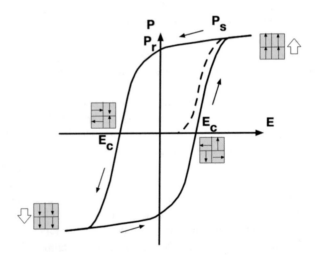

Fig. 21.6 Ferroelectric hysteresis loop. At E$_c$ the electrical dipoles are randomly aligned. With
increasing field strength, E, the dipoles become aligned until at P$_s$, they are all aligned and the
polarization is saturated. As the field is reduced the curve follows the arrow to the left. The material
retains a remenant polarization of P$_r$ when the field is zero. As the electric field is increased in the
opposite sense the dipoles again become randomized at E$_c$. As the field strength is increased further
the dipoles eventually become fully aligned in the opposite direction

capacitors in which the capacitance is tunable. They are used to make ferroelectric
random access memory for computers and in Radio Frequency IDentity (RFID)
chips in which small changes in voltage can be used to change the capacitance. In
addition to lead zirconate titanate, barium titanate (BaTiO$_3$) and lead titanate
(PbTiO$_3$) are common commercially-used as ferroelectric ceramics.

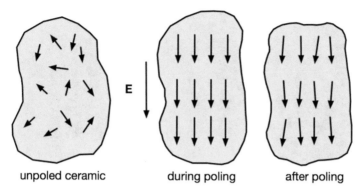

Fig. 21.7 On the left is a ceramic in which the dipoles in different regions of the material are aligned in different directions. The center schematic shows poling in which a large electric field is used to align the electric dipoles. On the right, after the electric field is removed, the dipoles remain largely aligned

PZT is made by mixing together powders of lead, zirconium and titanium oxides and heating to 800-1000 °C. The resulting powders are mixed with a binder and sintered to the desired shape. Upon cooling the crystal structure changes from cubic (Fig. 21.5a) to tetragonal (Fig. 21.5b). The different oriented crystals all have dipoles but since the crystals are randomly oriented there is no net dipole. In ordered to produce a permanent net dipole a large electric field is used to align the dipoles, a process called poling, see Fig. 21.7.

The market for piezoelectric materials was $16B in 2013 and this was expected to grow to at least $19B by 2019.

References

1. Callister, W. D., Jr. (2001). *Fundamentals of materials science and engineering*. New York: Wiley. ISBN: 0-471-39551-X.
2. Ho, J., Jow, T. R., & Boggs, S. (2010). Historical introduction to capacitor technology. *IEEE Electrical Insulation Magazine., 26*, 20.
3. Chung, C.-C. (2014). *Microstructural evolution in lead zirconate titanate (PZT) piezoelectric ceramics*, Ph.D. thesis. University of Connecticut.
4. Wang, J. J., Meng, F. Y., Ma, X. Q., Xu, M. X., & Chen, L. Q. (2010). "Lattice, elastic, polarization, and electrostrictive properties of $BaTiO_3$". from first-principles. *Journal of Applied Physics, 108*, 034107.

Chapter 22
Leather

When you think of leather you likely think of shoes, bags, belts, jackets and other clothing. The Assyrians, Ancient Egyptians, Ancient Greeks and Romans are all known to have used leather goods, clothes and footwear. Leather has even been used as part of armor. Animal skins, a by-product of killing animals for food, have likely been used for shoes, clothing, and tents for as long as mankind has been killing animals. Animal skins have also been used as parchment for writing since the Bronze Age[1]. But skins easily rot and smell if too wet, become stiff at low temperatures and can be damaged by higher temperatures.

The skin of a live animal, which is largely made up of interwoven, but moveable fibers of the protein collagen, has amazing properties: it is tough and hard-wearing, while being soft and flexible, and it is waterproof while allowing water vapor to escape. Dead skin does not share these properties: if allowed to dry out, it becomes brittle and hard, while it will putrefy if kept too wet. Thus, the aim of tanning an animal skin is to retain the outstanding properties of live skin, stabilize it and prevent its decay – as a spin-off, the process confers some resistance to degradation at elevated temperatures. The effect of tanning is that it prevents the collagen fibers in the skin from shrinking and sticking together and also lubricates them so that they can still move with respect to each other. Tanning might be considered the first time that humans used a natural material and processed it into a new material.

There is evidence both in the form of cave paintings and bone tools for scraping skins that indicate that leather was used in the Paleolithic period more than 40,000 years ago[2]. Mankind soon discovered that untreated animal skins were of limited use. Thus, methods of treating animal skins were developed, some probably by accident, such as: rubbing with animal fats; rubbing with brains, which provide emulsified oils; smoking, which provides formaldehyde from wood; and soaking in fermenting organic solutions followed by scraping off the hairs which have been loosened. Tanning using vegetable oils from tree barks (often oak; tannum is Latin

---

[1] http://www.leatherresource.com/history.html

[2] http://www.all-about-leather.co.uk/what-is-leather/where-does-leather-come-from.htm

© Springer International Publishing AG, part of Springer Nature 2018
I. Baker, *Fifty Materials That Make the World*,
https://doi.org/10.1007/978-3-319-78766-4_22

Fig. 22.1 A swatch of
leather made from a cow

for oak bark) and leaves, which provides tannin, a polyphenol molecule, dates back
at least to the fifth century Ancient Greeks. This process was continued until the
Industrial Revolution of the 18th and nineteenth century when demand for new
applications led to thinner, softer leather and eventually to the utilization of chromium salts for tanning in 1858.

Although at one time there would have been numerous oak-bark tanneries in the
U.K., there is but one left, J. and F.J. Baker & Co. Ltd. in Colyton, Devon[3]. This
tannery has been in the Baker family since 1862 and there has been a tannery on the
site since Roman times. The oak-bark leather takes 15 months to produce and is the
highest quality, most durable leather, and, no doubt, expensive. It is used for eques-
trian equipment and other leather goods such as paneling for rooms and shoes.

Modern tanning is a complex, multistep process that involves [1, 2]:

1. Soaking in water with additives to remove dirt, blood, salt etc.;
2. Liming and unhairing by using both sulfide and alkali solutions to loosen the
   keratin hairs, which can be removed mechanically or chemically;
3. Deliming and bateing, in which enzymes are added to remove any unwanted
   material such as epidermis that is left;
4. Pickling in typically formic or sulfuric acids to prepare the skin for tanning;
5. Tanning usually involves treating with chromium compounds to replace some of
   the collagen with chromium ions, which turns the hide blue;
6. Neutralizing, dyeing and fat liquoring involves neutralizing the acidity of the
   skin produced by the earlier processes, dyeing and then adding oils or fats that
   prevent the collagen fibers from sticking together, thus making the leather soft;
7. Drying, in which excess water is removed, by various means including vacuum
   drying;
8. Finishing, in which a surface coat is added to provide leather with an even color
   and texture.

65% of leather is made from the skin of cattle with most of the rest coming from
sheep (15%), goats (9%) and pigs (11%), see Fig. 22.1. However, leather is also
made from animals such as deer, ostrich, snake, kangaroo, crocodile, alligator, and
even salmon, eel and stingray [3]. Around half of all leather is used to make footwear, 14–15% is used for furniture, with around 10% each used for garments and
automotive applications, ~4–5% is used for gloves, with the rest used for all other

---

[3] http://www.jfjbaker.co.uk/about-us/

Fig. 22.2 About half of all leather is used to make shoes

applications,[4] see Fig. 22.2. Patent leather, which was invented in Europe but improved by the American Seth Boyden (1788–1870)[5], is often used for shoes. Originally patent leather was produced using various oil-based coatings such as linseed oil, but now the process normally uses various plastics to produce this flexible leather that is both shiny and waterproof [4].

Although leather is made from a natural product, there are significant environmental impacts from its production both in the wastewater from all the chemicals used in the processing, particularly the chromium salts, and in air pollution from various solvent vapors, since hydrogen sulfide and ammonia are released [5].[6] Although animal hides can be considered a by-product of meat production, a significant amount of a farmer's income can come from selling the hides, and for some animals, such as ostriches, the hides can be significantly more valuable than the meat. Thus, hide production is not nowadays simply a by-product, but contributes to the farming of animals, which itself has significant environmental consequences. For example, ~20% of greenhouse gas production arises from raising livestock [6].

World production of bovine hides is around 7 M tonnes (worth around $6B) with production in the developed world steadily decreasing and that in the developing world steadily increasing so that about 4 M tonnes are produced in the developing world.[4] China is the biggest producer of leather accounting for more than

---

[4] Food and Agriculture Organization of the United Nations, 2008.

[5] https://www.asme.org/engineering-topics/articles/manufacturing-processing/seth-boyden

[6] https://www.gizmodo.com.au/2014/06/how-leather-is-slowly-killing-the-people-and-places-that-make-it/

370 million sq. m. or around 25% of the total leather produced, more than twice the amount of the next biggest producer, Brazil, and ten times that of the U.S.A [6]. Interestingly, the ratio of the world population of cattle to the world population of people has remain fairly stable during the twentieth century ranging from 0.26 to 0.30, and the production of leather is closely related to the number of cattle [6]. Thus, as the population of the world continues to grow, the number of cattle will likely grow at a similar rate and the use of leather will continue to increase.

Even an ancient material like leather can be subject to innovation: transparent leather has recently been developed for fashionable clothing. Companies are also working on "growing" artificial leather using yeast genetically engineered to produce collagen, the main protein in leather [7].

References

1. The Chemistry of the Leather Industry, B.R. McMann and M.M. McMillan. New Zealand Institute of Chemistry, nzic.org.nz/ChemProcesses/animal/5C.pdf
2. Beghetto, V., Zancanaro, A., Scrivanti, A., Matteoli, U., & Pozza, G. (2013). The leather industry: A chemistry insight part I: an overview of the industrial process. *Sciences at Ca' Foscari*, 12–22. https://doi.org/10.7361/SciCF-448.
3. *Leather by Anna Ploszajski in Materials World*, published by the Institute of Materials, Minerals and Mining, Jan 2017, pp. 62–64.
4. Porter, R. (1850). Origin of malleable iron and patent leather. Scientific American, 5(46), 368.
5. Mwinyihija, M. (2010). Chapter 2. Main pollutants and environmental impacts of the tanning industry. In *Ecotoxicological diagnosis in the tanning industry*. NY: Springer Science+Business Media, LLC.
6. (2010). United Nations industrial development organization. In *Future trends in the World leather and leather products industry and trade, Vienna*.
7. More skin in the game. *The Economist*, August 26 2017.

Chapter 23
Lithium

Until the last few years, you probably hadn't heard of lithium apart from it being administered for bipolar disorder. Actually, the drug used is lithium carbonate. You can't feed someone elemental lithium.

You may remember lithium from high school chemistry where a favorite demonstration is to put lithium on water. Yes, on water. Lithium with an atomic number of 3 is the element with the lowest atomic number that is solid at room temperature and is the lightest elemental solid at only 534 kg m^{-3} and, thus, is much less dense than water at 1000 kg m^{-3}. The lithium reacts violently with water whizzing around the surface as it forms an oxide and releases hydrogen, which burns with a blue flame. The other alkali metals that one can do this with are sodium and potassium since they are also less dense than water at 968 kg m^{-3} and 862 kg m^{-3}, respectively, but their reaction with water is much more violent. Lithium is a silvery white metal that oxidizes so rapidly in air that it is stored under either argon or oil.

Lithium, from the Greek *lithos* or stone, was discovered by the Swedish chemist Johan August Arfwedson (1792–1841) in 1817 when analyzing the mineral petalite ($LiAlSi_4O_{10}$),[1] while the element itself was isolated four years later by the English chemist William Thomas Brande (1788–1866)[2] via electrolysis of lithium oxide (Li_2O).[3] Interestingly, current production methods also use electrolysis, but of lithium chloride.

Lithium has some interesting properties that underlie its use in some niche applications. Lithium has the highest specific heat capacity (a measure of the amount of energy required to raise the temperature of a unit mass of a material by 1 °C) at 3.58 kJ/kg.°C of all solids, only slightly less than that of water at 4.19 kJ/kg.°C, but significantly higher than that of ice at 2.11 kJ/kg.°C. Thus, it is useful as a heat transfer material because it can absorb large amounts of heat without its temperature rising significantly, and is, thus, used in heat transfer applications, such as in some

[1] https://www.thoughtco.com/johan-august-arfwedson-biography-607059
[2] https://www.revolvy.com/main/index.php?s=William%20Thomas%20Brande&item_type=topic
[3] It's Elemental – The Element Lithium.pdf.

© Springer International Publishing AG, part of Springer Nature 2018
I. Baker, *Fifty Materials That Make the World*,
https://doi.org/10.1007/978-3-319-78766-4_23

experimental nuclear reactors: the alkali metal sodium has also been used in the liquid state in some fast breeder nuclear reactors because of its low neutron absorption cross section, and both high thermal conductivity (142 W/m.$°$C) and high boiling point of 883 $°$C compared to water (0.591 W/m.$°$C, and 100 $°$C, respectively).

Lithium-6 (the isotope of lithium containing three neutrons and three electrons) was used in the first thermonuclear fusion or hydrogen bomb in 1952, designed by the Hungarian-American physicist Edward Teller (1908–2003) and Polish-American mathematician Stanisław Marcin Ulam (1909–1984) [1].[4] In this design, a nuclear fission bomb is surrounded by lithium deuteride (deuterium is an isotope of hydrogen with one proton and one neutron), in which the neutrons are absorbed and produce radioactive tritium (an isotope of hydrogen with one proton and two neutrons) and along with the high temperature produce nuclear fusion.[5] The details of the design are still secret, but are the basis of hydrogen bombs. Lithium's property of having a very low neutron cross-section (low likelihood of capturing neutrons) means that it is used in a number of different salts for cooling in nuclear power applications.[6]

An English chemist M. Stanley Wittingham (1941-) working at Exxon Research and Engineering, New Jersey, U.S.A. developed the first lithium ion battery, which had a titanium disulfide cathode and a lithium-aluminum anode, which Exxon manufactured. However, it was not a commercial success both because of the expense of the titanium disulfide and because of its reactivity. In modern Li-ion batteries, the positive electrode (cathode) is a lithium compound (for example lithium-doped cobalt oxide) and the negative electrode (anode) is a carbon-based material (for example graphite): lithium ions carry the charge from the anode to the cathode during discharging and in the opposite direction during charging. Both electrodes allow the lithium ions to move in (intercalation) and out (deintercalation) of them.[7]

According to data from the U.S. Geological survey,[8] in 1994 60% of lithium was used for ceramics, glass (lithium carbonate Li_2CO_3, added to a glass or ceramic makes them stronger) and primary aluminum production with other major uses being lubricants and greases and in synthetic rubber production. By 2006, the primary uses had not changed much - the batteries fell into the 'major other uses' category, see Fig. 23.1. By 2016, the picture had completely changed to 39% of lithium going into battery use, 30% glasses and ceramics, 8% into greases (lithium stearate-based greases are used for high temperature industrial, military, automotive, aircraft, and marine applications because they are non-corrosive and do not react with water or oxygen) and all other uses accounting for the other 25%. This growing use of lithium in batteries is reflected in the change of (non lead-acid) rechargeable batteries being used. In 1991, nearly 98% of rechargeable batteries were nickel-cadmium with nickel metal hydride making up the other portion [2]. In 1991, Sony

---

[4] http://www-groups.dcs.st-and.ac.uk/history/Biographies/Ulam.html

[5] https://www.britannica.com/technology/thermonuclear-bomb

[6] Lithium - World Nuclear Association.pdf.

[7] https://energyfactor.exxonmobil.com/science-technology/battery-changed-world/

[8] lithimcs96.pdf , lithimcs06.pdf, mcs-2017-lithi.pdf.

Fig. 23.1 A 4000 mAh 3.7
V lithium ion battery used
in a flashlight. The battery
is 64 mm long and 25 mm
in diameter, and replaces
three AA batteries in this
application

introduced the first lithium ion battery into their devices.[9] From then on the use of lithium in batteries took off so that by 1997 lithium ion and lithium polymer batteries represented 80% of the market and at that point the uses were in cell phones, cameras, laptops, power tools and medical equipment. The original electric and hybrid cars used lead acid batteries, but nickel metal hybrid batteries replaced most of these and were the dominant battery type through 2010. However, the fall in the cost of lithium ion batteries after 2005 meant that most newer hybrids and all electric vehicles, such as those produced by Tesla Motors, use lithium ion batteries because of the higher energy density (200+ W.h/kg) compared to nickel metal hybrid (30–80 W.h/kg) and lead-acid batteries (30–40 W.h/kg)[10]: Even so, lithium ion batteries only contain about one fiftieth the energy density of gasoline. The price per kWh of lithium ion batteries has fallen exponentially from around $1000 in 2008 to less than $145 in 2017 [3].

The revenue from Li-ion batteries is expected to increase to over $50 billion by 2020, from about $10 billion in 2009[11] with hybrid and electric vehicles accounting for two thirds of the market for Li-ion batteries. Nevertheless, in 2017 only around 1% of new cars are electric [3]. There is potential to increase the energy density of Li-ion batteries and intensive research is being conducted around the World to this end. In addition to their high energy density, the rate of loss of charge when not being used is less than half that of nickel metal hydride batteries and there is no memory effect. However, Li-ion batteries are not without issues: the charging and discharging has to be managed so that the battery is not degraded; they are sensitive to temperature, degrading at high temperature and don't work well at low temperatures; and completely discharging a battery can destroy it. The battery has to have good thermal management and can catch fire as witnessed by the notorious Samsung 7 phones catching fire and being banned from aircraft, and the fires on the Boeing

---

[9] BBC World Service podcast, Elements: Lithium, July 4th, 2014.

[10] http://www.epectec.com/batteries/cell-comparison.html

[11] Global Market for Lithium-Ion Batteries – Forecast, Trends and Opportunities 2014-2020, Taiyou Research, sample-8323376.pdf.

Fig. 23.2 Lithium whiskers can grow in lithium ion batteries. They can reduce battery efficiency, cause the batteries to short out, or possibly, to catch fire. They may have been the reason that the notorious Samsung 7 phones caught fire. (Courtesy of Weiyang (Fiona) Li)

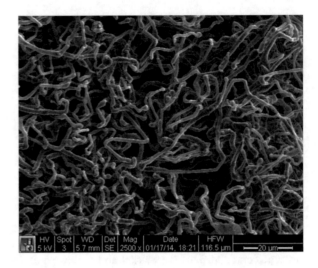

787 Dreamliner in 2013 due to the Li-ion batteries overheating.[12] These fires and overheating problems may have been due to the growth of Li dendrites (or whiskers) that can penetrate through the battery separator and thus cause a short circuit, see Fig. 23.2.

Lithium itself is very soft, but in recent years it has found its way into aluminum alloys for aerospace applications since the alloy's density decreases by 3% and the elastic modulus (stiffness) increases by 5% for every one weight percent of Li added to replace aluminum. Al-Li alloys can contain up to around 2.5 wt. % Li. The strength is also increased through both solid solution strengthening and by the formation of fine metastable Al_3Li precipitates, which both impede the motion of defects called dislocations by which the material deforms. These improvements come at a factor of three increase in cost.[13] A typical second-generation aluminum-lithium alloy such as 2099 contains 1.6–2 wt. % lithium along with 2.4–3.0 wt. % copper, 0.4–1.0 wt.% zinc and smaller quantities of other elements. This alloy can show a yield strength of 485 MPa with 10% elongation and a fracture toughness (the ability to withstand a crack propagating) up to 66 MPa.m$^{1/2}$. This compares to a similar yield strength of 520 MPa but a much lower fracture toughness of only 33 MPa.m$^{1/2}$ for alloy 7050 and a yield strength of much lower yield strength of 380 MPa but a fracture toughness of 53 MPa.m$^{1/2}$ for alloy 2026, both of which are non-lithium containing aluminum alloys used for aerospace applications.[14]

Mining of lithium mirrors the increase in lithium usage. In 1994, World production was 6100 tons, of which 2000 tons were produced in Chile. By 2006 World production had increased to 20,400 tons with Chile producing 8000 tons. Ten years later in 2016, World production had further increased to 35,000 tons, but now

---

[12] https://www.scientificamerican.com/article/how-lithium-ion-batteries-grounded-the-dreamliner/

[13] www.southampton.ac.uk/~jps7/.../manufacturing/aluminum-lithium.doc

[14] Smiths Metal Centres Ltd., Alloy Selection Sheet No. 34, 2099_SHP.pdf.

Australia is the biggest producer at 14,300 tons, although Chile's production over the ten years had also increased to 12,000, and Argentina, the third largest producer, mined 5700 tons. Australian production, most of which feeds the seemingly insatiable Chinese demand for raw materials, is by mining hard-rock spodumene, which is lithium aluminum inosilicate, $LiAl(SiO_3)_2$. In contrast, in Chile and producers such as Argentina, Bolivia, China, and the United States lithium is obtained from salt brine deposits. Sala de Attacama, the largest salt flat in Chile,[15] produces the World's lowest impurity content lithium and the lowest cost lithium since the salt is simply produced by evaporation over a period of up to two years. Although, maybe not for long. Salar de Uyuni in Chile's neighbor Bolivia is the world's largest salt flat, which at 10,582 square kilometers is more than three times the size of Sala de Attacama at 3000 km^2 and claims to contain 50–70% of the world's lithium reserves.[16] Bolivia started production in 2016. Lithium is 0.0017% of the Earth's crust with around 87% of the World's lithium reserves in brine deposits and only 13% in hard rock (see footnote 4).

References

1. Stwertka, A. (2012). *A guide to the elements*. New York: Oxford University Press. ISBN: 978-0-19-983252-1.
2. Goonan, T. G., (2012). *Lithium use in batteries*. Circular 1371. Reston: U.S. Geological Survey
3. Electrifying everything. *The Economist*, August 12th, 2017.

---

[15] http://www.sqm.com/en-us/acercadesqm/recursosnaturales/salmuera.aspx
[16] https://daily.jstor.org/salar-de-uyuni/

Chapter 24
Magnesium

You may have encountered magnesium not as the metal but as milk of magnesia, which is a suspension of magnesia or magnesium oxide ($Mg(OH)_2$) in water, in which it is insoluble. It is taken for various digestive tract ills. You may also have encountered it in medicinal Epsom salts, named after Epsom in Surrey, England, which is magnesium sulfate ($MgSO_4$). If you are a car enthusiast, you will have heard of Mag Alloy wheels. Magnesium alloys were the first materials to be used for die cast wheels since they are very lightweight compared to steel, and, thus, have a high specific strength and a high damping capacity[1]. They were first produced in the 1930s but were largely superseded by aluminum wheels in the 1960s apart from in the competitive racing market. If you are reading this in a car or on a plane, you may well be sitting on a seat whose frame is made of magnesium.

Magnesium is a silvery white metal, whose name is derived from Magnesia, a district of Thessaly, Greece. Magnesium was discovered by the Scottish physician and chemist Joseph Black (1728–1799)[2]. It was a one of the elements, along with potassium, sodium, calcium, strontium, barium, and boron, that was isolated by Sir Humphrey Davy (1778–1829), who used electrolysis in 1808 to separate it from its oxide [1].[3] At 2.1%, it is the eighth most common element in the Earth's crust: at a concentration of 1300 p.p.m. in seawater (the second highest cation content after sodium), there is essentially an infinite supply. Commercially, magnesium is made either from ores of magnesite (magnesium carbonate, $MgCO_3$) or dolomite (calcium magnesium carbonate, $CaMg(CO_3)_2$), or from naturally-occurring salt deposits called salt brines that contain around 10% magnesium chloride.[4] Magnesium is produced from its ores by thermal reduction, which involves heating to produce magnesia (MgO) and the use of ferrosilicon to reduce the magnesia to magnesium. Outside China, magnesium is mostly produced from salt brines via electrolysis.

---

[1] https://www.carsdirect.com/aftermarket-parts/pros-and-cons-of-magnesium-wheels

[2] http://www.chem.gla.ac.uk/~alanc/dept/black.htm

[3] http://www.bbc.co.uk/history/historic_figures/davy_humphrey.shtml

[4] https://www.thebalance.com/magnesium-production-2339718

© Springer International Publishing AG, part of Springer Nature 2018
I. Baker, *Fifty Materials That Make the World*,
https://doi.org/10.1007/978-3-319-78766-4_24

Fig. 24.1 Close packed hexagonal crystal structure of magnesium. The ratio of the length of the axes is 1.62, which is very close to the ideal packing of a hard spheres in a hexagonal array of 1.63

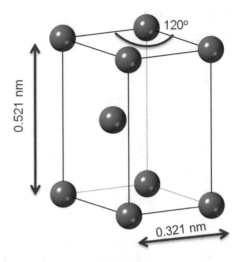

0.521 nm

120°

0.321 nm

Magnesium is very reactive and highly flammable as a powder. It was used as a powder when mixed with potassium chlorate to produce white light for photography and is used in fireworks and flares since it produces a brilliant white light burning at 3100 °C. Magnesium wire was at one time used in single-use flashbulbs that contained pure oxygen and were electrically triggered.

Bulk magnesium is less reactive since it forms protective oxide and nitride films. Magnox alloys, magnesium alloys containing aluminum and small amounts of other metals, whose name is derived from **MAG**nesium **No OX**idation were used to clad uranium in the first commercial electricity-producing power stations called Magnox reactors. The first of these reactors was at Calder Hall, England in 1956 although it was more useful for producing plutonium for nuclear weapons than at producing electricity since it had a very low efficiency of 18.8%. While these alloys were used because of their low neutron capture cross-section, the alloys suffer from the fact that they react with water. While corrosion resistance is good in air, it is poor in seawater.

Magnesium and its alloys are not as strong as structural alloys based on aluminum, iron or nickel with typical yield strength of wrought alloys of 160–275 MPa and ultimate tensile strengths of 180–440 MPa. They also have less ductility at room temperature since the hexagonal crystal structure does not provide many deformation modes, see Fig. 24.1. Magnesium alloys have moderate corrosion resistance, although the addition of aluminum helps, but they are poor in marine environments. However, one characteristic of magnesium alloys makes them very desirable, particularly in transportation applications: they are very lightweight. The density of magnesium at 1740 kg/m^3 is only two thirds that of aluminum at 2700 kg/m^3 and one fifth that of steel at 8050 kg/m^3. Thus, they have a very good strength-to-weight ratio, often called the specific strength, of 158 kN.m. kg^{-1}, better than aluminum alloys (115 kN.m. kg^{-1}) or low-carbon steels (46 kN.m. kg^{-1}). While magnesium is more expensive than aluminum per unit weight ($2.23/kg for magnesium versus

Fig. 24.2 Transmission
electron microscope image
showing MgZn$_2$
precipitates (dark features)
in a rolled and annealed
ZM61 magnesium
(Mg – 6 wt. % Zn -1 wt.%
Mn) alloy. (Courtesy of
Min Song)

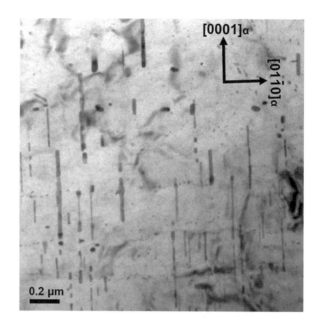

Fig. 24.3 A high-
resolution transmission
electron microscope image
of a twin boundary. In
ZM61 magnesium
(Mg – 6 wt. % Zn -1 wt.%
Mn) alloy. The lines of
atoms on either side of the
twin boundary, which runs
diagonally from top left to
bottom right, are mirrored
across the boundary.
(Courtesy of Min Song)

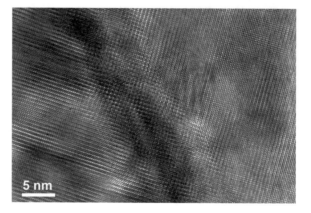

$1.93/kg for aluminum[5]), if the two materials are compared on a volume basis, magnesium is cheaper than aluminum ($3880/m^3 for magnesium versus $5200/m^3 for aluminum). A disadvantage of magnesium alloys is the low modulus (stiffness) of 45 GPa particularly when compared to steel at 200 GPa. Magnesium alloys are also limited by its relatively poor strength at high temperatures and combustibility.

Typical magnesium alloys contain aluminum, zinc, manganese, silicon, zirconium and/or copper, see Figs. 24.2 and 24.3. One of the strongest alloys ZK60A, which has 5 Wt. % Zn and 0.45 wt. % Zr has a yield strength of 285 MPa and 11% elongation. The relatively low ductility of magnesium alloys means that parts are

---

[5] http://www.infomine.com/investment/metal-prices/aluminum/

either cast or hot-worked at 200–350 °C, where the alloys are more ductile. In addition to their use in automobile seats and increasingly aircraft seats, magnesium alloys are used in steering wheels and columns, car dashboards, power tools, lawn mower casings, racing bikes, transmission cases, laptops and cellphones, missiles and airplane parts – magnesium was first used in airplane parts in World War I. A major use of magnesium is as an alloying agent in aluminum and iron alloys. Magnesium is also used as a sacrificial or galvanic anode that is connected to an iron structure to form a circuit and preferentially corrodes to protect the steel.

Magnesium is the third most commonly used metal behind iron and aluminum, but even so as of 2013 magnesium alloy use was only 1 million tons compared to 50 million tons of aluminum. China produces 85% of World's magnesium.[6] The use of magnesium will continue to grow strongly in transportation applications, replacing some plastics in some structural applications since it is stiffer, more recyclable and less costly to make. Future possible uses include in magnesium ion batteries, and bioabsorbable non-toxic implants.

References

1. Chaline, E. (2012). *Fifty minerals that changed the course of history*. Buffalo: Firefly Books Ltd.

---

[6] USGS, Mineral Commodity Summaries, Magnesium, January 2018.

Chapter 25
Nickel-Based Superalloys

The reason that vast numbers of people fly these days is because flying is relatively inexpensive. One factor in this low cost is what every flyer knows, that the amount of space per passenger is becoming less and less, and so more people are crammed into the same space. But once the aircraft has been paid for, the big operating cost for airlines is the cost of the fuel, and fuel efficiency has increased by over 45% since the 1940s [1]. Aircraft have become increasingly fuel efficient by changing from turbojet engines, where all the air from the compressor blades (the large blades that you can see at the front of the engine) passes through the engine, to turbofans, where most of the air traveling through the compressor blades passes around the engine, with increasingly higher bypass ratios (the amount of air that goes around the engine compared to the amount of air that passes through the engine). An essential step is to run the engine hotter, which improves efficiency. The key to doing this was the introduction and development of nickel-based superalloys used for the turbine blades in the hottest part of the engine.

The first patent for a jet engine was granted in 1930 to Sir Frank Whittle (1907–1996), who was a pilot officer in the Royal Air Force at the time [2].[1] The first two demonstration jet engines both ran successfully in 1937. One, the hydrogen-burning Heinkel HeS 1,[2] was designed by Hans von Ohain (1911–1998) in Germany while the other, the kerosene-fueled Power Jets WU, was built by a company co-founded by Whittle. The latter engine technology was transferred to the U.S. Air Force and General Electric for further development [3]. Turbojet engines were used in order to produce fighter aircraft with higher speeds than the fastest turboprop (propeller) driven plane. These two engines led to the first two fighter jet prototypes, the German Heinkel He 178, which first flew in 1939, and the British Gloster E.28/39, which first flew in 1941. The Messerschmitt Me 262, which first flew in July, 1942, became the first operational jet fighter aircraft in April, 1944.[3] It used the alloy steel

---

[1] https://www.britannica.com/biography/Frank-Whittle

[2] https://www.britannica.com/biography/Hans-Joachim-Pabst-von-Ohain

[3] https://www.space.com/16650-first-fighter-jet.html

© Springer International Publishing AG, part of Springer Nature 2018
I. Baker, *Fifty Materials That Make the World*,
https://doi.org/10.1007/978-3-319-78766-4_25

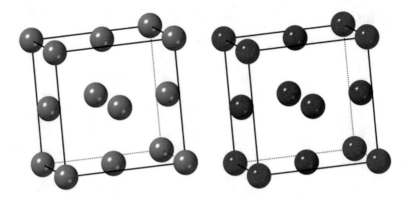

Fig. 25.1 Schematic of the f.c.c. (left) and ordered f.c.c. or L1$_2$ (right) crystal structures that occur in nickel-based superalloys. In the f.c.c. structure different kinds of atoms can sit in any location in the unit cell, whereas in the L1$_2$ structure the nickel atoms occupy the positions in the middle of the faces and other atoms occupy the corners of the unit cell. The unit cells in both cases are around 0.36 nm

4340 for its turbine blades, which had to be replaced every 5 h since this material was really not suitable for the purpose.

These early turbofan engines had to use the available materials of the time with the turbine blades made of corrosion and oxidation-resistant stainless steels. Whittle's engine initially used "Stayblade" steel, a face-centered cubic austenitic stainless steel, see Fig. 25.1, containing 20 wt.% chromium and 8.5 wt.% nickel with smaller amounts of manganese, silicon, titanium and carbon, for both the turbine blades and turbine discs. However, due to the material's inadequate strength, it was soon replaced in the turbine blades by Firth-Vickers Rex 78 steel [4, 5], which contains 14 wt.% chromium and 18 wt.% Ni also with small amounts of manganese, silicon, titanium and carbon and in the turbine discs by G18B steel, which contains 13 wt.% chromium, 13 wt.% Ni, 10 wt. % cobalt, and smaller amounts of niobium, tungsten, molybdenum, manganese, silicon and carbon [6]. However, because of their low strength at high temperatures, stainless steels limited the operating temperature to around 800 °C [7]. The need for better materials that can operate at higher temperatures led to the invention of nickel-based superalloys in the 1940s and their subsequent development as Nimonic alloys in the U.K., Tinidur alloys in Germany, and Inconel alloys in the U.S.A. [1]. The chemical and microstructural development of superalloys would probably have amounted to little without the introduction of vacuum induction melting (VIM), which enabled alloys to be cast that no longer suffered from the low ductility due to defects and lack of cleanliness that occurred in air cast materials. VIM improved both the alloy's strength and temperature capability.

The creep (the continuous elongation of a material under load when at high temperature) resistance of nickel-based superalloys arises from the presence of coherent ordered f.c.c. Ni$_3$(Al,Ti) precipitates in the f.c.c. matrix, see Fig. 25.2. Figure 25.1 is schematic of the crystal structures of the two phases, which have very similar

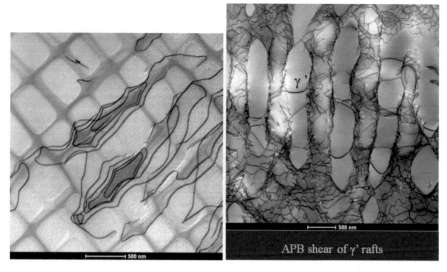

Fig. 25.2 Transmission electron micrographs of the microstructure of a Ni-based superalloy showing $L1_2$ precipitates in a f.c.c. matrix. The dark lines are linear defects called dislocations. (Courtesy of P.R. Subramanian)

lattice parameters and so fit together well, that is, they are said to be coherent. The precipitates, which are very stable at high temperature, are often referred to as γ' (gamma prime) and the matrix as γ (gamma). The precipitate's role is to impede the movement of linear defects called dislocations through the material, see Fig. 25.2. The use of the γ' precipitates has also evolved over time. In order to improve the creep properties further some current alloys having very high volume fractions of 60–70% of the γ' and some use duplex heat treatments, where microstructures consisting of large cuboidal γ' precipitates with fine γ' precipitates between them are produced.

The introduction of nickel-based superalloys immediately led to increases in the operating temperatures of jet engines. Since their introduction, metallurgists have strived to improve these materials through improvements in both their chemistry and microstructure. The first nickel-based superalloys were polycrystals, see Fig. 25.3. The grain boundaries between the crystals or grains are weak points in the material for creep (time dependent deformation under load), fatigue (repeated loading) and oxidation. Grain boundary carbides, such as $Cr_{23}C_6$, were added to the earlier superalloys to improve the grain boundary strength.

Later, directional solidification was used to grow turbine blades from the molten alloy with columnar grains along their length, see Fig. 25.3. In this case, the grain boundaries are all aligned along the length of the turbine blade, which is the direction of tensile load on the blade. The columnar-grains prevent grains or crystals sliding past each other under load, so-called "grain boundary sliding" since in order to slide past each other the grains have to be at an angle to the direction of the load on the blade. Directionally-solidified turbine blades were first introduced by Pratt

Fig. 25.3 Photograph of turbine blades consisting of (from left to right) polycrystals, columnar grains, and a single crystal

and Whitney Aircraft (P&WA) for the twin-engined General Dynamics F111 fighter and were first used in the 1970s in engines introduced by P&WA for commercial aircraft such as the Boeing 747 and McDonnell Douglas DC-10 [8]. Eventually, single crystals turbine blades were developed, which completely eliminated all grain boundaries, with the crystals oriented in a particular direction, see Fig. 25.3. This also led to better resistance to creep, fatigue and oxidation. Single crystal blades were first introduced by P&WA in the Boeing 747, 757 and 767 and Airbus A310 in 1980 [4]. All large commercial aircraft and all military aircraft now use single crystal turbine blades. The use of a columnar grain structure allowed an increase in material operating temperature of 40 °C while the use of single crystals increased the operating temperature by a further 40 °C (see footnote 3). Figure 25.4 shows how the innovations in both material, microstructural control and chemistry impacted the operating temperature as a function of time.

While huge advances have been made to increase the operating temperatures of Ni-based superalloys through the use of single crystals and improved chemistry, further substantial improvements are unlikely through this route. However, engineering innovations such as hollow blades through which cooling air is blown to form a protective air shroud around the blade, which allow the blade to operate above its melting point and the use thermal barrier coatings to prevent oxidation of the blades have continued to push the operating temperature envelope to around 1750 °C, see Fig. 25.5. To continue to improve engine efficiency with current engine designs, higher temperature materials need to be developed. What these will be is unclear at the moment, but some candidates include multi-phase alloys containing silicides or ceramic composites.

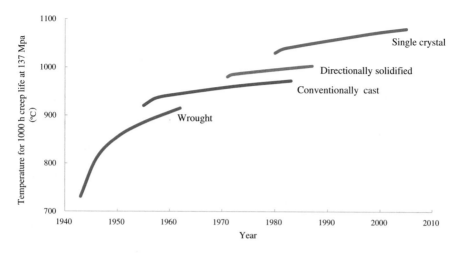

Fig. 25.4 Material operating temperature in gas turbine engines versus time

Fig. 25.5 A G.E.
superalloy blade with
cooling holes through
which air is blown that
forms an air shroud around
the blade. (Courtesy of
P.R. Subramanian)

References

1. Schafrik, R. Materials in jet engines: past, present and future. academia.edu/4751249/Materials_in_Jet_Engines_Past_Present_and_Future P26-63.
2. "Sir Frank Whittle, O. M., K. B. E.. 1 June 1907–9 August 1996" by G. B. R. Feilden and W. Hawthorne, *Biographical memoirs of the Royal Society*. http://rsbm.royalsocietypublishing.org/content/44/435, Published 1 November 1998. doi:https://doi.org/10.1098/rsbm.1998.0028
3. Sims, C. T. (1984). A history of superalloy metallurgy for superalloy metallurgists. In M. Gill, C. S. Kortovich, R. H. Bricknell, W. B. Kent, & J. F. Radavich (Eds.), *Superalloys* (pp. 399–419). Warrendale, PA: The Metallurgical Society of the American Institute of Mining, Metallurgical and Petroleum Engineers.
4. Pavelec, S. M. (2007). *The jet race and the second world war*. Annapolis: Naval Institute Press. ISBN 978-1-59114-666-7.
5. (February 1963). Flying steel: 25 years of stainless steel in the jet age. *FLIGHT International, 21*, 261.
6. Oakes, G., & Barraclough, K. C. (1981). In G. W. Meetham (Ed.), "Steels" in The development of gas turbine materials (pp. 31–62). London: Applied Science Publishers, Ltd. ISBN-13: 978-94-009-8113-3.
7. Koff, B. L. (2004). Gas turbine technology evolution: A designers perspective. *Journal of Propulsion and Power, 20*(4), 577–595.
8. Gell, M., Duhl, D. N., & Giamei, A. F. (1980). *The development of single crystal Superalloy turbine blades. Superalloys*. In: Proceedings from the 4th International Symposium on Superalloys, Americal Society of Metals, Metals Park, OH. pp. 205–214. ISBN 0871701022.

Chapter 26
Nitinol

Nitinol is the most prominent and most utilized of a class of seemingly magical materials called Shape Memory Alloys that can "remember" a shape from a different temperature or at a different stress. The shape memory effect was discovered by a Swedish Chemist Arne Ölander (1902–1984) in gold-cadmium alloys in 1932. The effect was later observed in a beta brass (Cu-Zn) alloy in the 1950s and in Nitinol in 1959. The name Nitinol arises because the shape memory effect was discovered in a nickel titanium alloy by William J. Buehler and Frederick Wang at the Naval Ordnance Laboratory. Hence, the name **Ni**ckel **Ti**tanium **N**aval **O**rdnance **L**aboratory [1, 2].

The first practical use of a shape memory alloy, nearly 40 years after the discovery of the shape memory effect, took place in 1970 when a tube of Cryofit, a nitinol alloy, was used for couplings in a high pressure hydraulic system on a U.S. Navy F-14 fighter [3]. Nowadays, many shape memory alloys are known, but 70–75% of the applications use nitinol, 20–25% use Cu-Al alloys and less than 5% of applications use other alloys [4]. The Cu-Al alloys are used less because they have a smaller shape memory effect and are not as corrosion resistant. However, they are less expensive. The key market driver is the medical sector which utilizes 62% of nitinol, but nitinol is also used in the aerospace (5%), automotive (3%), and robotics (4%) industries, with all other uses accounting for 26% in 2013. Common uses for nitinol is for stents to open up arteries, as eyeglass frames and as "dental brass", see Fig. 26.1 [5].

The unusual properties of nitinol arise from the two different crystal structures that it adopts around room temperature and at slightly high temperatures. Around room temperature nitinol has the monoclinic crystal structure shown in Fig. 26.2, which is referred to as a B19' martensite. This structure deforms by a process called deformation twinning in which planes of atoms shuffle past each other when a high enough stress is applied. At higher temperature, nitinol adopts a much simpler, ordered-body-centered cubic or B2 structure that is called austenite, see Fig. 26.2.

If the martensite crystal structure is heated it starts to transform to austenite at a temperature referred to as A_s and is fully austenite at a temperature A_f, see Fig. 26.3.

© Springer International Publishing AG, part of Springer Nature 2018
I. Baker, *Fifty Materials That Make the World*,
https://doi.org/10.1007/978-3-319-78766-4_26

Fig. 26.1 A nitinol archwire used to adjust the position of teeth. (Courtesy of D.W. Van Citters)

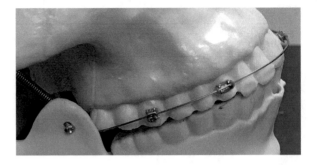

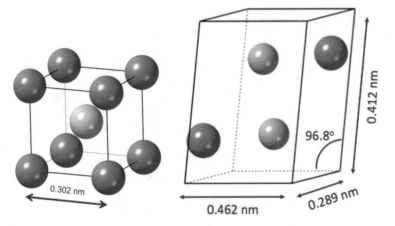

Fig. 26.2 The crystal structures of NiTi. Left: Austenite, which is ordered body-centered cubic; Right: Martensite, which adopts the B19' monoclinic crystal structure [9]

If the austenite crystal structure is cooled it starts to transform to martensite at a temperature M_s and is fully martensite at a temperature M_f. Thus, there is thermal hysteresis in the transformation. This hysteresis loop width can range from 20–50 °C depending on the exact composition of the nitinol. The M_s and A_s temperatures change very rapidly for small changes in the Ni:Ti ratio, see Fig. 26.4. Both the size of the loop and the temperatures can be changed by alloying with either elements. Importantly, the transformation is instantaneous and reversible.

The so-called one-way shape memory effect in nitinol can be illustrated with a two-dimensional model, see Fig. 26.5. Let us start with the material above the temperature A_f so that it is fully austenite, see top image in Fig. 26.5. If the material is cooled below the temperature M_f it fully forms the twinned martensite structure shown bottom left in Fig. 26.5. If the twinned martensite is deformed it changes into the detwinned martensite shown right in Fig. 26.5. Upon unloading the material retains this shape change. However, when the detwinned structure is again heated above the temperature A_f, it again changes to austenite and to its original shape. This is the thermal shape memory effect. This effect is used in stents where a collapsed stent is inserted into an artery, which upon warming expands to its original shape

Fig. 26.3 Schematic showing how the martensite to austenite and the reverse transformation occurs in nitinol at different temperatures

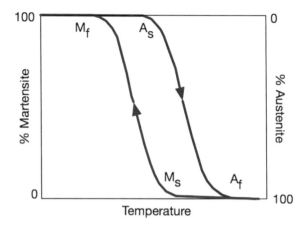

Fig. 26.4 Graph showing how the martensite start temperature for the change from austenite to martensite in nitinol changes with nickel content

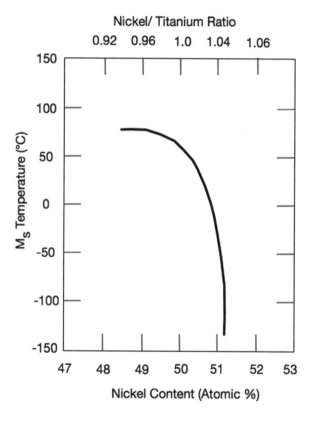

and opens up the artery. The effect is also used in dental wires and brackets, where a wire is placed in the mouth at lower temperature and upon warming it changes shape producing a force on teeth and causing them to move. The nitinol alloys used for this have to be austenite at body temperature. Even though nitinol contains

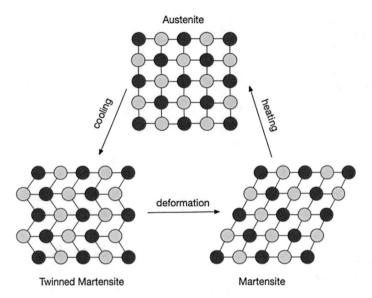

Fig. 26.5   2-D schematic showing how the Shape Memory Effect operates. If austenite is cooled it forms the twinned martensite shown left. Upon deformation under a load the martensite becomes untwined and remains so even if the load is removed. Upon heating, the martensite returns to the Austenite structure and returns to its original shape [10]

Fig. 26.6   Comparison of a stress-strain curve for a steel with that of nitinol in the austenite state. The steel only shows elastic behavior up to the point E (around 0.1%) beyond which it is permanently deformed. Nitinol can show pseudoelasticity up to the point labeled P, which can be 9% [5]

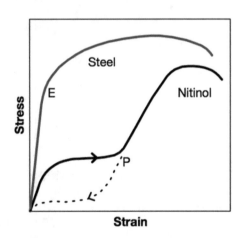

nickel, which may not be good for the body, nitinol can be made biocompatible by electro-polishing and forming a passivating titanium dioxide layer on the surface, which both prevents corrosion and the release of nickel ions. Nitinols can also show a two-way shape memory effect by "training" the alloy, but this effect is little used since the change of shape is typically half that in the one-way shape memory effect.

Another useful feature of nitinols and other shape memory alloys is called superelasticity or pseudoelasticity. A typical metallic alloy such as steel can be stretched by 0.1–0.2% until it reaches its elastic limit, which is beyond the point it will be permanently deformed, see Fig. 26.6. This is because at low strains the bonds in the crystal lattice are simply being stretched. In contrast, nitinol can be stretched from 4% to 9% depending on the composition and temperature, see Fig. 26.6 [6]. This behavior occurs at temperatures above A_f, that is when the material is fully austenite. If stress is applied to the austenite the change of shape that occurs is accommodated by transforming to martensite. If the stress is removed, the martensite changes back to austenite and the original shape is resumed. This behavior only occurs up to a temperature called M_d above which the austenite behaves like a normal metal. This superelastic effect is exploited in eyeglass frames and provides a comfortable fit and resistance to damage.

Nitinol is quite difficult to process. It is made by either vacuum arc melting or vacuum induction melting followed by hot working. Cold working is problematic because of the pseudoelasticity. Machining is also problematic because of the poor thermal conductivity (0.18 W/cm.°C for austenite and 0.086 W/cm.°C for martensite).

As noted earlier, nitinol is used in many industries and applications. It is used for couplings, biomedical applications, toys, actuators and sensors [7, 8]. It is used as an actuator in many applications, one of which is in showers where it can be used to shut off water flow if the temperature of the water increases above, say, 45 °C. Other applications include as underwires in brassieres and, of course, as with all high-tech materials it has been used in golf clubs. In the latter case, the pseudoelasticity means that the ball can remain in contact with the club head for long times thus imparting more energy and more spin. The first biomedical applications of nitinol were in the late 1980s as a device to localize breast tumors and as a bone anchor in orthopedic surgery.

By 2014, there were over 20,000 patents worldwide on shape memory alloys and their applications. As of 2013 the market for nitinol was $3.1 B and this was expected to have increased to $8.1 B by 2018. The biomedical applications seem boundless and uses are found in almost every part of the body. Similarly, aircraft uses range from parts of the engine to the ailerons and landing gear, and in automobiles from the engine to the brakes and structural parts. The uses of nitinol are expected to increase tremendously in the future.

References

1. Kumar, P. K., & Lagoudas, D. C. (2008). Introduction to shape memory alloys. In D. C. Lagoudas (Ed.), *Shape memory alloys* (pp. 1–25). New York, NY: Springer Science and Business Media.
2. Morgan, N. B. (2004). Medical shape memory alloy applications – The market and its products. *Materials Science and Engineering A, 378*, 16–23.
3. Wu, M. H., & Schetky, L. M. D. (2000). Industrial applications for shape memory alloys. In *Proceedings of the international conference on shape memory and superelastic technologies* (pp. 171–182). Pacific Grove, CA.

4. McWilliams, A. (2015). *Smart materials and their applications: Technologies and global markets, AVM023E*, BCC Research, Wellesley, MA.
5. Van Humbeeck, J. (2015). *Shape memory alloys for biomedical application*. esomat, Antwerp, Belgium, September 14–18, 2015.
6. Mertmann, M. (2004). Non-medical application of Nitinol. *Minimally Invasive Therapy & Allied Technologies, 13*(2), 254–260.
7. *Applications for shape memory alloys*. https://depts.washington.edu/matseed/mse_resources/Webpage/Memory%20metals/applications_for_shape_memory_al.htm
8. Jani, J. M., Leary, M., Subic, A., & Gibson, M. A. (2014). A review of shape memory alloy research, applications and opportunities. *Materials and Design, 56*, 1078–1113.
9. Otsuka, K., & Ren, X. (2005). Physical metallurgy of Ti-Ni-based shape memory alloys. *Progress in Materials Science, 50*, 511–678.
10. Kapoor, D. (2017). Nitinol for medical applications: A brief introduction to the properties and processing of Nickel Titanium shape memory alloys and their use in stents. *Johnson Matthey Technology Review, 61*, 66–76.

Chapter 27
Nylon

At one time "nylons" was synonymous with "stockings", but the terminology has fallen out of use as, to a large extent, as has the wearing of stockings. Nylons were introduced as an affordable alternative to silk stockings and were an instant hit. Nylon was invented at E. I. du Pont de Nemours and Company, Wilmington, DE, U.S.A. by the American chemist Wallace Carothers (1896–1937) in 1935 [1–3] and in 1939, the year that Nylon stockings first went on sale, 64 million pairs were sold[1]. This use declined rapidly to zero along with other commercial uses of Nylon, such as making toothbrush bristles, during the Second World War as Nylon was diverted to the U.S. war effort to make tire cords, flak vests, ropes and parachutes. The latter application arose when Asian supplies of silk that had been used to make parachutes dried up [1].

Nylon is actually the generic name for a range of synthetic thermoplastic polymers – polymers that can be melted and reset repeatedly - called polyamides, so called because they have repeating amide -CO-NH_2 groups, see Fig. 27.1. Nylon 6, 6, or polyamide 6, 6, which was invented by Carothers, accounts for half of all nylons produced, see Fig. 27.2. Nylon 6,6 is produced through reacting two identical molecules of HOOC-$(CH_2)_4$-COOH each of which contain six carbons:

$$n\text{HOOC} \ (\mathbf{CH_2})_4 \ \text{COOH} + n\text{H}_2\text{N} \ (\mathbf{CH_2})_6 \ \text{NH}_2 \rightarrow \left[OC \ (\mathbf{CH_2})_4 \ CO \ \text{NH} \ (\mathbf{CH_2})_6 \ \text{NH} \right]_n + 2n\text{H}_2\text{O}$$

The production of Nylon 6, 6 was the first use of a condensation polymerization reaction to make a polymer, in which a small molecule, which for Nylon production is water (H_2O), is lost. (This is as opposed to "addition polymerization" where molecules simply join together without the loss of a small molecule.)

Nylon 6,6 has a glass transition temperature (the temperature at which a polymer reversibly changes from a hard, glassy material at lower temperatures to a soft, rubbery material at higher temperatures) of 49 °C and a melting point of 243-260 °C and can be extruded, granulated or made into fibers. Replacing the methyl groups,

[1] http://www.smithsonianmag.com/arts-culture/why-Nylon-run-over-180954954/

© Springer International Publishing AG, part of Springer Nature 2018
I. Baker, *Fifty Materials That Make the World*,
https://doi.org/10.1007/978-3-319-78766-4_27

Fig. 27.1 The amide   all nylons
group that is the basis of

$$
\begin{array}{cc}
O & H \\
\| & | \\
- C - N -
\end{array}
$$

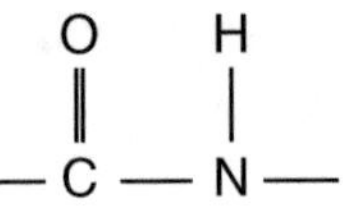

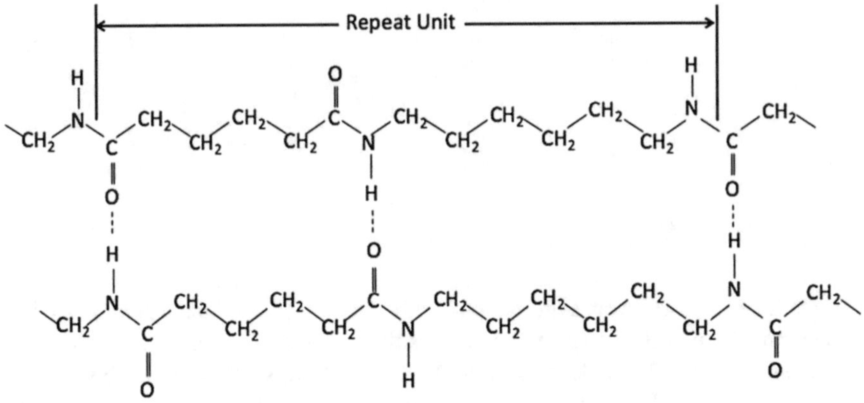

Fig. 27.2 Nylon 6,6 or polyamide 6,6. The repeating unit consists of two sets of six carbon atoms. Hence the name. The regular structure enables the molecules to line up to form lamellar crystals when the polymer is slowly cooled from the melt. The individual polymer chains are joined together by hydrogen bonds (indicated as dotted lines) between the oxygen and hydrogen atoms

CH_2, (bolded in the above reaction) with other organic chemical groups leads to a range of different nylons with different properties, for example Nylon 6,12, Nylon 4,6, Nylon 6, and Nylon 12.

Above the melting point Nylon behaves like a viscous liquid with randomly-arranged polymer chains. When it solidifies, much of the polymer forms lamellar crystals in which the molecules are aligned while the remaining regions are amorphous, that is it is essentially a frozen liquid state. Nylon 6, 6 is highly crystalline because of the regular nature of the polymer and the crystalline regions are held together with hydrogen bonds between the polar amide groups ($-CO - NH-$) groups, see Fig. 27.2. This crystalline plus amorphous character, like many polymers, is the basis of the properties of the material: the crystalline regions provide rigidity and strength, while the amorphous regions improve the elasticity.

The properties of Nylons vary widely depending on the particular Nylon, but they all have excellent properties including good strength. For example, they exhibit yield strengths of 45–83 MPa and ultimate tensile strengths of over 95 MPa [4][2]; by comparison the tensile strength of the strongest polyethylene, HDPE, is 30 MPa. They also show good ductility (15–300% elongation) toughness, abrasion resistance, and a low coefficient of friction. These excellent mechanical properties lead to the use of nylons for technical applications like bearings, gears, cams and bushings. They are also used in a wide variety of consumer applications including

---

[2] http://www.bpf.co.uk/plastipedia/polymers/Polyamides.aspx

Fig. 27.3 Some
applications of Nylon as
stockings, for toothbrush
bristles and for a comb

zippers, fabric, stockings, carpets, handles, combs, cooking bags and fishing line, see Fig. 27.3. Nylons are used also for coatings on cables and wires because of the high electrical resistance of 1×10^{13} Ohm.m and high dielectric breakdown strength (the electric field strength at which an insulator is no longer electrically insulating) of 2000–3000 kV/m.

One downside of nylons is that because of the polar amide groups present, see Figs. 27.1 and 27.2, they absorb water and other polar liquids and swell. The amount of water absorption and, hence, the amount of swell depends on the particular Nylon. Unlike the most common polymer, polythene, Nylons are heavier than water (1000 kg/m^3) with a density of 1130–1350 kg/m^3.

While nylons have some very good properties, they are about five to six times the price of polyethylene, the most common polymer. Thus, Nylons or polyamides as a whole account for only about 1% of all polymers used. Even so, their use is expected to grow at about 5.5% annually to a market size of $30 billion by 2020 [1] both because of the general increase in demand for polymers and because of the high strength coupled with good ductility that make them the ideal materials for some applications.

References

1. Amato, I. (1998). *Stuff: The materials the world is made of*. New York: Avon Books, Inc ISBN-10: 0380731533.
2. Strong, A. B. (2000). *Plastics, materials and processing* (2nd ed.). Upper Saddle River: Prentice Hall ISBN: 0-13-021626-7.
3. van Dulken, S. (2000). *Inventing the 20th century: 100 inventions that shaped the world. From the airplane to the zipper*. New York: New York University Press ISBN: 0-8147-8808-4.
4. Callister, W. D. (2001). *Fundamentals of materials science and engineering*. New York: Wiley, Inc ISBN: 0-471-39551-X.

Chapter 28
Paper

With the advent of personal computers, the paperless society was predicted by the British-American information scientist Frederick Wilfrid Lancaster (1933–2013) [1]. As is evident in the graph in Fig. 28.1, it didn't quite work out that way and the amount of paper consumed continues to increase every year. In fact since Lancaster's prediction in 1978 the amount of paper and paperboard used has roughly tripled with much of the growth coming from developing economies.

Paper is one of the four great inventions of Ancient China, which are celebrated in Chinese culture. One of the others, printing, is also clearly related to paper – the other two are gunpowder and the compass. The invention of paper was a disruptive technology, replacing bone, bamboo slips or scrolls, which were heavy, or silk, which was expensive. The fascinating Changsha Bamboo Slips Museum, which I visited in 2015, was established in 2002 to house many of the 140,000 bamboo slips, wooden slips and wooden tablets dating from the Wu Kingdom (222–280 AD) that were discovered in central Changsha, Hunan Province, China in 1996: later discoveries at this site found bamboo slips dating back to the second century B.C. West Han period.[1]

The invention of paper is typically attributed to a courtier to the Han dynasty named either Cai Lun or Ts'ai-Lun (~48–121) around 105 AD although paper appears to have been produced at least a hundred years before this date.[2] The technology that he developed involved mixing fibers of materials such as rags, hemp waste, fishing net, mulberry and other plant fibers in water, beating them to a sludge, lifting the suspended fibers out of the water on a screen thereby draining the water, and then drying into a thin matted sheet, a process that is fundamentally the same as that used today, although nowadays mostly wood fibers are used and the process is more sophisticated and mechanized.

After it spread throughout China, papermaking then diffused to South East Asia (Japan, Korea, Indo-China), and paper mills were established in Samarkand and

---

[1] http://china.org.cn/english/culture/153360.htm

[2] http://www.hqpapermaker.com/paper-history/

© Springer International Publishing AG, part of Springer Nature 2018
I. Baker, *Fifty Materials That Make the World*,
https://doi.org/10.1007/978-3-319-78766-4_28

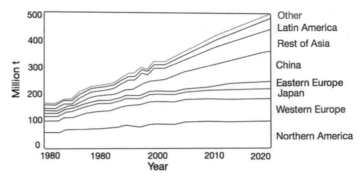

Fig. 28.1 Global demand for paper and paperboard from 1980 to 2020. (http://blog.fluid-eng.com/2013/08/paper-an-indispensable-material-for-2000-years/)

Baghdad by the end of the eighth century although the Arab world knew of paper and used it before then [2]. Paper was used in Africa by the ninth century, replacing the use of papyrus [3]. Papermaking probably arrived in Europe in the eleventh century either via Spain, introduced by the occupying Moors from North Africa, or, possibly, through Italy. It gradually spread north, but it was not until 1490 that the first paper mill in England was established.

Paper was not, of course, the first material used for writing. The Sumerians likely invented writing around 4000 B.C. using soft clay tablets that were later hardened in sunlight. Papyrus, from which we derive the word paper, was developed by the Ancient Egyptians in the fourth century B.C. It is made by pounding the sticky fibrous material in the stem of the papyrus plant, a reedy plant that grows in shallow water, into thin sheets. Parchment, made from animal skins, including Vellum from calf-skin, was used by the Ancient Greeks. Writing on stone, metal sheets, bark or leaves was used throughout various parts of the world including in the Americas before the arrival of Europeans by at least the fifth century AD.

The book, originally a codex, was invented by the Romans although the first books were not made of paper – they may have been made of papyrus or vellum. However, paper is ideal to be made into a book since it is very thin and hundreds of pages can be bound together representing a huge repository of knowledge in a compact form.

Before the industrialization of papermaking, paper was uncommon and expensive. The Bible printed by the goldsmith Johannes Gutenberg (1398–1468) in 1454/55 using mechanical movable type can be thought of as the beginning mass communication although at the time a book was still far beyond the means of most people to buy. Even so it would have been far cheaper than making the bible from parchment, which it has been estimated would require the skins of 300 sheep.[3] The reduction in cost eventually came about through the industrialization of paper making from using water-powered mills, improved machinery, and eventually the switch from making paper starting with rags of cotton, hemp and linen (flax) to

---

[3] http://conservatree.org/learn/Papermaking/History.shtml

Fig. 28.2 Note the right hand side of the paper has started to yellow after lying for only 1 day exposed to the New Hampshire summer sun

making paper from wood - some paper is still made from cotton for example for paper banknotes and high-end art paper. The invention of cheap paper and printing had profound effects allowing ideas to be shared and, eventually, to widespread literacy.

The improvements in wood pulp production from which paper is made ranged from the stamp mill to the Hollander Beater [4], the invention of wood-grinding machines in 1844 simultaneously by the German Friedrich Gottlob Keller (1816–1895) and the Canadian Charles Fenerty (1821–1892) and the conical or Jordan refiner, by the American Joseph Jordan in 1858.

The goal of different chemical, thermomechanical (heating wood chips in pressurized water) and grinding processes for making wood pulp is to separate the cellulose fibers in wood. The first two processes remove the lignin (delignification) that holds the cellulose fibers together, while the later does not. Such processing is not required for making paper from cotton since it is already 90% cellulose. Not removing the lignin produces a higher yield and cheaper paper, but also creates paper that yellows over time as the lignin reacts with oxygen when in sunlight, see Fig. 28.2. Once the wood pulp, which is cellulose fibers, in water is obtained it can simply be pressed and dried, see Fig. 28.3. Of course, papermaking is a bit more complicated

Fig. 28.3 X-ray micro
computed tomography
image showing the fibers
in paper. (Courtesy of
M.K. Kynoch and
R.W. Obbard)

than that. The paper is usually bleached to make it white; clay or chalk (calcium carbonate) are added to improve the paper's attributes for writing or printing; and sizing is added to the surface to make ink dry on the surface rather than be absorbed into the paper. Many different materials have been used for sizing starting in the eighth century with starch in China and the Arab world.[4] Later glue and gelatin, both of which were derived from animals, were used. From about 1850 rosin, which is derived from pine tress, was used in mass-produced paper.[5] Unfortunately, rosin also leads to the degradation of the paper. Paper may also be coated to have matte or gloss finishes for printing, wax can be added to make wax paper so that it is moisture-resistant and non-stick and it can be coated with light-sensitive chemicals to turn it into photographic paper amongst other uses.

Paper production has a significant environmental impact, with 35% of all logged trees being used to produce paper, although many of those trees are replanted by paper companies. Paper production also uses large amounts of chemicals that can be environmental contaminants. In addition to paper made from wood pulp, a significant amount of paper is made from recycled paper. While recycling of paper reduces the need for logging significantly[6], it doesn't necessarily reduce the chemical or energy usage in making paper. Paper recycling varies by region with the highest rates of around 70% in Europe.[6]

---

[4] http://cool.conservation-us.org/coolaic/sg/bpg/annual/v05/bp05-11.html

[5] https://www.naturalpigments.com/art-supply-education/sizing-paper-gelatin/

[6] http://www.paperrecycles.org/statistics

Lancaster's prediction of a paperless society has not come about – maybe it never will. Paper has more uses than as a medium for the written word. It is used for bags, photographic paper, wrapping material, packing material such as cardboard and as waxed paper, money and toilet paper. Long term, it is not clear that paper will continued to be used for money as we become a cashless society and even bank notes are made of plastic, which is more expensive but lasts much longer and incorporates many safety features. It has been used for the latter purpose since at least the sixth century AD in China.[7] The cheapness of paper along with the invention of printing not only led to the production of books but also to the daily newspaper, which of course is now dying out. Paper is also finding new uses as an alternative to plastic packaging, and environmentally-friendly coatings are being developed for paper that are not based on petroleum-based hydrocarbons.

References

1. Lancaster, F. W. (1978). *Toward paperless information systems*. New York: Academic.
2. Needham, J., & Tsuen-Hsuin, T. (1985). *Science and civilization in China: Chemistry and chemical* (p. 5). Cambridge, UK: Cambridge University Press.
3. Amato, I. (1998). *Stuff: The materials the world is made of*. New York: Avon Books, Inc ISBN-10: 0380731533.
4. Biermann, C. J. (1996). *Handbook of pulping and papermaking* (2nd ed.). Cambridge, MA: Academic Press. ISBN: 978-0-12-097362-0.

[7] http://encyclopedia.toiletpaperworld.com/toilet-paper-history/complete-historical-timeline

Chapter 29
Platinum

Platinum is not a common element at $3.7\% \times 10^{-7}\%$ or 0.003 parts per million (p.p.m.) in the Earth's Crust, although it is about ten times more common than gold. It is in a group of noble precious metals in the middle of the transition metal block of the Periodic Table referred to as the platinum group metals, which includes ruthenium, rhodium, palladium, osmium and iridium. Platinum is a silvery white metal, whose name is derived from the Spanish for silver *plata*. Platinum, along with the platinum group metals rhodium, palladium and iridium adopts the face centered cubic crystal structure, see Fig. 29.1 – the other two platinum group metals ruthenium and osmium adopt the hexagonal-close-packed crystal structure. Platinum is one of the heaviest elements with a density of 21,500 kg m^{-3} – the heaviest element is another platinum group metal osmium at 22,590 kg m^{-3}. Its most useful properties are that it is very malleable and, thus, can be drawn into wire and beaten in thin sheets, and its excellent corrosion and oxidation resistance.

The Briton, Charles Wood (1702–1774), first isolated platinum in 1741. Nevertheless several other Europeans could also claim to have first recognized platinum as an element. However, South American natives had been mining and using platinum (often mixed with rhodium and palladium) to make jewelry for at least 2000 years [1], and the oldest known piece of platinum was used as a coating on a case, found in Thebes, Egypt dating to the time of the seventh century B.C. Egyptian Queen Shapenapit [2]. These ancient platinum items were made by beating naturally-occurring platinum or platinum alloys since the melting point of platinum at 1775 °C is much too high for these ancient civilizations to have produced.

Very small amounts of platinum occur naturally, but it is usually extracted from the ores sperrylite ($PtAs_2$) and cooperite (Pt, Pd, Ni)S using a flotation process. These ores, along with a number of other ores than contain other platinum group metals, are found in South Africa and some other sites. In Canada and Russia, platinum and other platinum group metals are by-products of the extraction of nickel and copper by electro-refining. Processing either involves chemical processing or heating with air blown through. About 30% of platinum comes from recycling. In 2016, 120,000 kg of the world total of 172,000 kg were produced in mines in

© Springer International Publishing AG, part of Springer Nature 2018
I. Baker, *Fifty Materials That Make the World*,
https://doi.org/10.1007/978-3-319-78766-4_29

Fig. 29.1 The face centered cubic structure adopted by platinum. The side of the unit cell is 0.3924 nm

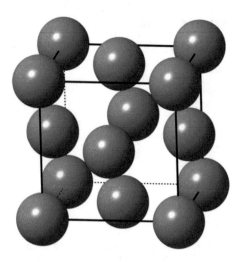

South Africa, far greater than the next largest producer Russia at 23,000 kg. Indeed of the known reserves of platinum group metals of 67 million kg, 63 million kg are in South Africa.[1] Note that platinum's price, like that of gold, is quoted in troy ounces, with 1 kg equaling 32.15 troy ounces.

Platinum has had an important use for setting standards in the metric system. From 1889 to 1960, the meter was defined as the distance between two lines scribed on a platinum – 10% iridium bar which was held at 0 °C at the International Bureau of Weights and Measures (BIPM) near Paris. This was later replaced by a definition based on specific number of wavelengths of light. Since 1889 the kilogram has been defined by the mass of a platinum-10% iridium bar at BIPM. Platinum use in these applications relied on its oxidation resistance and high density. The iridium addition substantially increased the hardness of the platinum making it less likely to scratch.

Platinum is used in many industries including oil refining, dentistry, the chemical industry, and electrical and electronics industries. Applications include in spark plugs, in computer hard disks, as a part of a thermocouple (Fig. 29.2), furnace windings, in optical fibers, protective coatings on missiles and fuel nozzles in jet engines, as an electrode in fuel cells and in heart pacemakers The similarity of its thermal expansion coefficient of 8.8×10^{-6} °C^{-1} to glass allows its use as an electrical contact in glass systems. It is used in some anti-cancer drugs such as cisplatin. However, its biggest use by far is as a catalyst. It is used as a catalyst in the production of many chemicals including silicone, nitric acid, benzene, and in oil refining. However, over 50% of its use is in catalytic convertors in vehicles, which convert carbon monoxide to carbon dioxide; unburnt hydrocarbons to carbon dioxide and water; and nitrogen oxides to nitrogen. Another large use of platinum is in jewelry, see Fig. 29.3. For example, the British Imperial State Crown is made of gold, platinum and silver, while the crown of George VI's wife Queen Elizabeth was made wholly from platinum.

---

[1] https://minerals.usgs.gov/minerals/pubs/commodity/platinum/mcs-2017-plati.pdf

Fig. 29.2 Platinum-5%
rhodium 0.254 mm
diameter thermocouple
wire

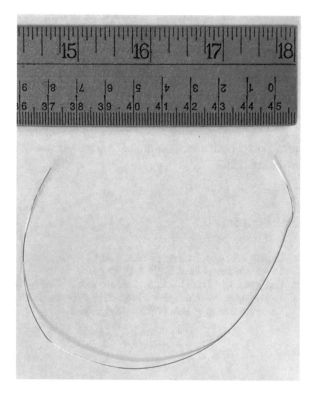

Fig. 29.3 Men's platinum
rings. (Courtesy of Julie
Von Bargen Thom, Von
Bargen's Jewelry, Hanover,
N.H., U.S.A)

Platinum coins have also been produced by various countries such as the U.S.A. and
the U.K. as investment media. Largely because of its use as a catalyst in producing
many chemicals, it has been estimated that one in five products either incorporate or
use platinum in their manufacture [1]. Because of its high value, platinum is extensively recycled from automotive and electronics applications, and recycled platinum
satisfies around 25% of the global demand for platinum [3].

Platinum accounts for around 51% of the use of platinum group metals with palladium and rhodium accounting for a little under 24% each. The latter two metals
have similar applications to platinum and are largely used as catalysts with again the

largest use in automotive exhausts. They are both less common in the Earth's Crust than platinum (0.003 p.p.m.) at 0.006 p.p.m. and 0.0002 p.p.m., respectively. This is presumably one of the reasons that palladium and rhodium are both more expensive than platinum ($988/troy oz) at $1104/troy oz. and $1690/troy oz.

At the beginning of 2018, platinum is only slightly cheaper than gold at $988/troy oz. compared to gold at $1334/troy oz.. The price has been quite volatile since 1990 ranging from a low of $372/troy oz. in 1998 to a high of $1721/troy oz. in 2011.[2] Most of this volatility is driven by the volatility in the vehicle market. The price is expected to increase around 4% per year until 2023 due to a modest increase in demand [3].

References

1. Emsley, J. (2001). *Platinum, Nature's building blocks: An A-Z guide to the elements*. Oxford: Oxford University Press ISBN: 0-19-850340-7.
2. Gladstone, J. H. (1901). Berthelot and the metals of antiquity. *Nature, 65*, 82–83.
3. Subramanian, V. (November 2014). *"Precious metal: Global markets", AVM098A*. BCC Research, Wellesley, MA. ISBN: 1-56965-996-6.

---

[2] https://www.lme.com/Metals/Precious-metals/Platinum

Chapter 30
Polyester

Most likely you hadn't considered that the clothes that you are wearing are made out of the same material as the bottle that your fizzy drink arrives in or that your expensive coffee beans are stored in. The clothing label will indicate polyester or a cotton/polyester mix, the bottle material will likely be called PET and the bag will be shiny aluminized mylar, but they are all forms of polyester.

Unlike some other polymers such as polyethylene and polystyrene, polyester is a not a single polymer but a family of polymers. The first useful polyester was developed in 1941 by two British chemists, John Rex Whinfield (1901–1966) and James Tennant Dickson,[1] while working for a British textile company with the peculiar name of The Calico Printers' Association of Manchester.[2] These researchers reacted terephthalic acid (International Union of Pure and Applied Chemisty (IUPAC) name: Benzene-1,4-dicarboxylic acid) with ethylene glycol (IUPAC name: ethane-1,2-diol) to produce polyethylene terephthalate (IUPAC name: Poly(ethyl benzene-1,4-dicarboxylate)), which is commonly known as PET, see Fig. 30.1. This process is a so-called condensation reaction in which the condensate (the molecule that is left over) is water (H_2O). Not only was this the first polyester produced but it is also the one that is produced in the highest volume today. The British company Imperial Chemical Industries (ICI) developed the polymer into a patented synthetic textile fiber and marketed it as Terylene in 1949.[3] ICI and the American company E. I. du Pont de Nemours and Company had a research development and patent sharing agreement at the time, and in 1945 DuPont purchased the U.S. development rights to polyester. By 1950, DuPont had produced a strong, wrinkle-resistant polyester fiber, that it called "Dacron" for clothing [1]. Dupont subsequently developed

[1] Polyester Film History, History of PET, What is Polyester Film, PET, How is Polyester Film made, PET.pdf

[2] http://manchesterhistory.net/manchester/tours/tour6/stjames.html

[3] https://blog.oureducation.in/properties-and-use-of-terylene/

© Springer International Publishing AG, part of Springer Nature 2018
I. Baker, *Fifty Materials That Make the World*,
https://doi.org/10.1007/978-3-319-78766-4_30

Fig. 30.1 The reactants
terephthalic acid and
ethylene glycol are used to
make the polymer
polyethylene terephthalate
or PET. Note the reaction
is more complicated than
shown and requires a
catalyst

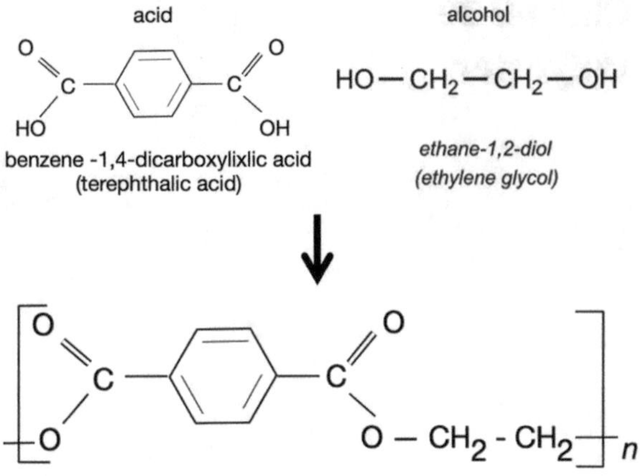

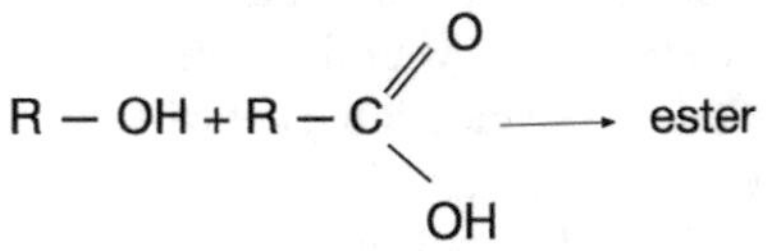

acid   alcohol

benzene -1,4-dicarboxylixlic acid
(terephthalic acid)

ethane-1,2-diol
(ethylene glycol)

Fig. 30.2 The general reaction of a carboxylic acid (molecule on the left) with an alcohol (molecule in the center) to produce a polyester, where R is a hydrocarbon group

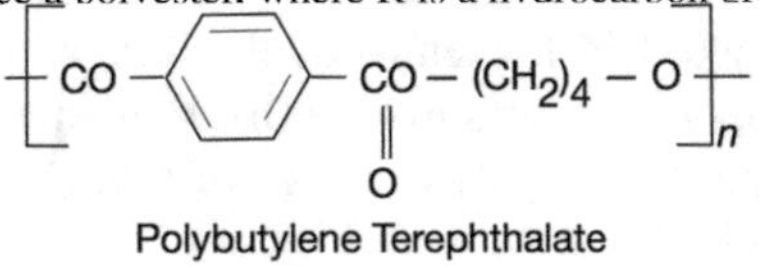

Polybutylene Terephthalate

Fig. 30.3 The poly(oxy-1,4-butanediyloxycarbonyl-1,4-phenylenecarbonyl) or PBT monomer

a polyester film called Mylar in 1952.[4] It took until 1973 before a DuPont employee
Nathaniel Wyeth (1911–1990) was granted a patent for a making a bottle out of
PET, thus, revolutionizing the bottle industry: the PET bottle is lightweight, recyclable, a good gas and liquid barrier, strong and impact resistant.

For different types of applications the PET fibers are arranged differently: for
fibers the molecules are mainly arranged in one direction; in films they are in two
directions; and for packaging they are in three directions.

As noted earlier, polyester is a family of polymers. The general reaction is of a
carboxylic acid with an alcohol and is of the general form shown in Fig. 30.2.[5] The
second most commonly used polyester is polybutylene terephthalate (IUPAC name:
Poly(oxy-1,4-butanediyloxycarbonyl-1,4-phenylenecarbonyl)) referred to as PBT,
see Fig. 30.3. This is produced from terephthalic acid and 1,4 butanediol. The latter
is like ethylene glycol, but has two more CH_2 groups. Each additional CH_2 group in

[4] http://www.polyesterconverters.com/pcl_apps/stage1/stage2/applications_and_enduses/historyofpet.htm

[5] The Structure of Polyester.pdf

the alcohol used to make a polyester lowers the melting point. PBT is a (semi-) crystalline thermoplastic engineering polymer – engineering plastics have either better mechanical or thermal properties (sometimes both) than other plastics and are usually more expensive. It is often used as an insulator in the electrical and electronics industries. PBT is strong and heat-resistant up to 150 °C (or up to 200 °C if reinforced with glass fibers) and resistant to solvents. In comparison with PET, PBT has slightly lower strength and stiffness, but has slightly better impact resistance and has low moisture absorption, good fatigue resistance and good self-lubrication behavior. It also has a slightly lower density (1300–1500 kg/m^3 versus 1400–1600 kg/m^3). PBT is easier to mold and crystallizes more rapidly than PET because it has a glass transition temperature (the temperature when a material changes from being hard and glassy material to soft and rubbery) of 170 °C compared to 240 °C for PET. PBT also melts at a lower temperature of 230 °C versus 260 °C for PET. In fact, care must be exercised to ensure that molded PET products become fully crystallized since the partially crystallized portions may lead to cracking, crazing and shrinkage. Another common non-engineering polyester is poly-1,4-cyclohexylene-dimethylene terephthalate or PCDT. This is weaker than PET, but its greater elasticity and resiliency enables its use in items such as furniture coverings and drapes.

While PET is produced as small granules, it can be melted and squeezed through fine holes and the resulting filaments spun into fibers. Polyesters fibers are widely used in clothing, alone or in blends with other fibers, usually cotton. Its strength and resistance to wear are utilized in cords for car tires, conveyor belts and hoses. Polyester fibers have a number of desirable attributes for clothing: they are very strong (PET has about twice the strength of high density polyethylene, HDPE, at up to 80 MPa), a good modulus of elasticity (the resistance to stretching) for a polymer (2–4 GPa, compared to less than 1.2 GPa for HDPE), it is quite ductile (elongations up to 125% are possible), durable, resistant to most chemicals, resistant to shrinking, wrinkle resistant (and so are often blended into cotton clothes), mildew resistant, abrasion resistant, and they are hydrophobic and, hence, dry quickly. Unfortunately, polyester fibers are released every time clothes are washed and end up in the environment (as do other fibers from clothes) [2] and the dyes associated with dyeing with polyester are not very environmentally friendly. However, polyester can be 100% recycled: it has a resin identification code of "1". It can also be degraded by two different bacteria.[6] Although recycled PET is not often used in food containers, it is finding growing use in carpets. It is even being considered for applications such as reinforcing concrete [3].

In 2015, PET accounted for 8.6% of the non-engineering plastics global market of 229 million tons, and the polyesters (PET and PBT) accounted for 6.4% of the global engineering plastics market of 25 million tons; PET accounting for 7% of the total plastics market.[7] The annual World-wide production of PET was approximately 40 million tonnes and is growing at 7% per year. Of this, about 65% is used to make fibers, 5% for film and 30% for packaging.

---

[6] http://www.pslc.ws/macrog/pet.htm
[7] The Plastic Industry Berlin Aug 2016 - Copy.pdf

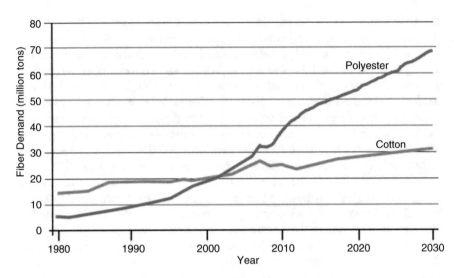

Fig. 30.4 Demand for cotton and polyester fibers. (After http://www.textileworld.com/textile-world/fiber-world/2015/02/man-made-fibers-continue-to-grow/)

Fig. 30.5 The omnipresent PET bottle

From the moment PET fibers were first produced, demand for them grew. In 2014 polyester fibers accounted for 83% of the synthetic fiber market, with polyamide a distant second at 8% polyacrylics third at 3.5%, and all other synthetic fibers making up 6.5% of the market. These proportions are not expected to change significantly in the near future. The production of all synthetic fibers was around 52 million tonnes in 2014 and this is expected to grow with an annual growth rate of around 10% to over 85 million tonnes by 2019.[8] The leading producers are China, India and the U.S.A. Around 2000, PET fibers overtook the production of cotton fibers as the most used textile fibers. Around 2010 the polyester fibers were used more than all other textile fibers combined, and by 2030 PET fiber production is projected to be more than double that of cotton,[9] see Fig. 30.4. PET bottles also seems to be everywhere, Fig. 30.5.

References

1. Strong, A. B. (2000). *Plastics, materials and processing* (2nd ed.). Upper Saddle River: Prentice Hall ISBN: 0-13-021626-7.
2. Napperan, I. E., & Thompson, R. C. (2016). Release of synthetic microplastic plastic fibres from domestic washing machines: Effects of fabric type and washing conditions. *Marine Pollution Bulletin, 112*, 39–45.
3. Pelisser, F., Montedo, O. R. K., Gleize, P. J. P., & Roman, H. R. (2012). Mechanical properties of recycled PET fibers in concrete. *Materials Research, 15*(4), 679–686.

---

[8] 2015-2019 Synthetic Fibers Market, technavio insights

[9] http://www.textileworld.com/textile-world/fiber-world/2015/02/man-made-fibers-continue-to-grow/

Chapter 31
Polyethylene

Polyethylene, also called by its International Union of Pure and Applied Chemistry (IUPAC) systematic name polyethene, is commonly known as polythene. It is a thermoplastic polymer meaning that it can be melted (at 120–180 °C) so that it flows and can be cooled to the solid state repeatedly. The name polyethylene is derived from the fact that the molecule consists of many (poly) "mers" or repeat units of the gaseous molecule ethylene (Fig. 31.1). It is the simplest possible carbon-based polymer consisting only of a backbone of carbon atoms with hydrogen atoms attached.

Ethylene is converted into polyethylene, see Fig. 31.1, by heating to 150–300 °C at a high pressure of 100–300 MPa (1000–3000 atmospheres) in the presence of very small quantities (<10 parts per million) of oxygen that acts as a reaction initiator. The chains vary both in length and in how branched they are, both of which depend on the method of production (Fig. 31.2). Greater branching decreases a variety of properties including density, melting point, strength, toughness, transparency, permeability and resistance to solvents [1]. Upon cooling from the molten state, some of the molecules can become aligned and produce crystalline regions (Fig. 31.2). The degree of crystallinity of a polymer depends both on the length of the individual molecules and their branching: shorter molecules and less branching lead to greater crystallinity. Low-density polyethylene (LDPE) and high-density polyethylene (HDPE) have degrees of crystallinity of 45–55% and 70–80%, respectively. Stretching or rolling the polymer increases the alignment of the polyethylene molecules. Crystalline polyethylene has a higher density, higher strength and greater resistance to chemical attack, but it is brittle and becomes opaque since the boundaries between crystals reflect the light.

Polyethylene was first synthesized by the German chemist Hans von Pechmann (1850–1902) in 1898, who produced it inadvertently when heating diazomethane, which is a yellow gas at room temperature.[1] It was "rediscovered" by Eric Fawcett (1902–1983) and Reginald Gibson (1908–1987) in 1933 while working at Imperial

---

[1] https://www.britannica.com/topic/industrial-polymers-468698/Polyethylene-PE#ref608626

© Springer International Publishing AG, part of Springer Nature 2018   163
I. Baker, *Fifty Materials That Make the World*,
https://doi.org/10.1007/978-3-319-78766-4_31

Fig. 31.1 The ethylene
molecule, a mer or repeat
unit that makes up the
polyethylene molecule, and
part of a polyethylene
molecule

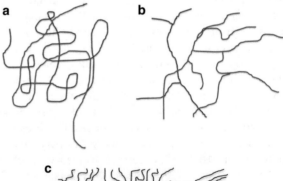

Fig. 31.2 Schematics
showing (**a**) largely linear
polyethylene molecules,
which pack together well
as in HDPE, (**b**)
polyethylene molecules
showing many branches,
which pack together more
loosely as in LDPE, and
(**c**) how polyethylene
molecules are aligned in
some regions to produce
crystals and amorphous in
other regions

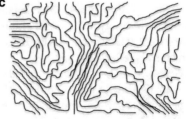

Fig. 31.3 SPI resin ID code for high-density polyethylene from a container

Chemical Industries (ICI) in England [2].[2] They produced a waxy material when they heated the gas ethylene with the more complex organic molecule benzaldehyde to a temperature of 170 °C at a very high pressure of 193 MPa (1900 atmospheres). Michael Perrin (1905–1988) and John Paton, also at ICI, later developed this into the first practically useful polyethylene synthesis,[3] and the first polyethylene plant was opened by ICI in 1939. The first uses of polyethylene were in airborne radar equipment and submarine cables because of its low density (slightly lower density than water) and excellent properties as insulator – the resistivity of polyethylene at around 1×10^{15} Ohm-m is huge compared to the value for a metal such as copper of 1.7×10^{-8} Ohm-m. Production of polyethylene from ethylene nowadays is somewhat more sophisticated and always involves catalysts since the ethylene molecule from which it is produced is quite stable.

Polyethylene is typically produced as granules that can be mixed with dye – which I did for my first summer job at age 15 – and then extruded. A similar processing route is used for other thermoplastic polymers such as polypropylene.

The molecular weight of an individual molecule of ethylene is 28 g/mole, while the molecular weight of polyethylene depends on how many mers make up the chain, which can be from 1000–10,000. There are several types of polyethylene but most of the polyethylene used is either LDPE (resin identification code, which used to be called a recycling symbol, 4) or HDPE (resin identification code 2), see Fig. 31.3. The difference in density between these two types of polyethylene is not that large: LDPE is typically 910–940 kg/m^3, while HDPE is 930–970 kg/m^3: both are less dense than water (1000 kg/m^3). This difference is derived from the greater branching of LDPE of about 2% of the carbon atoms present, which results in

---

[2] http://www.edn.com/electronics-blogs/edn-moments/4410771/Polyethylene-synthesis-is-discovered--again--by-accident--March-27--1933

[3] http://www.icis.com/resources/news/2008/05/12/9122447/polyethylene-discovered-by-accident-75-years-ago/

weaker forces between the molecules, producing a polymer with lower strength. The lower level of branching and greater packing density for HDPE compared to LDPE translates into a higher ultimate tensile strength (30 MPa versus 10 MPa), and a higher elastic modulus (the ability to resist stretching) of ~1 GPa versus ~0.25 GPa, properties which depend on the exact processing conditions. Both LDPE and HDPE are weaker and have a lower modulus than other common plastics such as ABS, polypropylene, PVC, PET, polystyrene and Nylon.[4] Conversely, both LDPE and HDPE have very low glass transition temperatures (the temperature at which a polymer changes from ductile and rubbery to glassy), of around -100 °C, compared to other polymers and so can be used at low temperatures without becoming brittle. HDPE is more opaque, and can withstand slightly higher temperatures (say 120 °C rather than 95 °C) for short periods than LDPE, but both can exhibit elongations to failure of several 100%.

Polyethylene is, by far, the most used polymer, accounting for about one third of all plastics used with uses too numerous to mention but for a few. LDPE is used for various containers, bottles, tubing, films, and various molded laboratory equipment. Its greatest use, by far, is for plastic bags. The greater strength of HDPE and good strength to weight ratio means that it is used in more demanding structural applications with one-third of its use being for milk jugs and other hollow containers. It is also used for plastic lumber, piping, chairs and tables, fuel tanks, bottles, and in various applications where its electrical insulating properties are useful.

Two other common types of polyethylene are linear low-density polyethylene (LLDPE) and ultra high molecular weight polyethylene (UHMWPE), both of which are less dense than HDPE. LLDPE (density ~ 915 kg/m^3) consists of mostly linear, but short polymer chains, which tend not to pack together well. It is easier to process for some applications, where it can be used to make thinner and, hence, cheaper material for applications such as plastic bags, plastic wrap, toys, covers, pipes, buckets and containers, cable covering and flexible tubing. It is displacing more traditional HDPE and LDPE in many applications.

At the other end of the scale, UHMWPE molecules consist of 200,000 mers in contrast to the molecules that make up HDPE, which consist of 700–1800 mers. These very long chains are hard to pack together giving a density of only 930–935 kg/m^3, but the long chains make a material that is very tough and has excellent chemical resistance (the ends of the chains are easier to attack) and wear resistance with a low coefficient of friction. It is also stronger than HDPE with a strength of 40–50 MPa but with an elastic modulus between that of HDPE and LDPE of ~0.55–0.7 GPa. Hence, UHMWPE is a critical component in various biomedical applications such as knee or hip replacements where its wear resistance properties enable its use to separate the metal components that slide on it, see Fig. 31.4.

Polyethylene is resistant to attack by many chemicals, but it is susceptible to UV radiation, which breaks the bonds in the polymer chain and causes it to become brittle, a problem that can be solved at low cost by adding 1–3% carbon black, which absorbs the UV energy.

---

[4] http://www.matweb.com/reference/tensilestrength.aspx

Fig. 31.4 Hip implant. The white hemispherical part is made of UHMWPE. Courtesy D.W. Van Citters

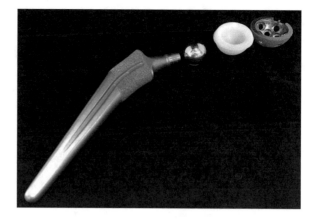

While the future demand for polyethylene is expected to be flat in Europe and to grow slowly in North America, demand is expected to continue increasing rapidly in the rapidly-growing economies of Asia, particularly China and India, where the market size is already bigger than that of Europe and North America combined at ~40 million tonnes.[5] Brazil is the third biggest producer. The Global market size in 2013 was $127 billion and 81 million tonnes and is expected to grow to $152 billion and 94 million tonnes by 2018, a steady compound annual growth rate in revenue of 3.7%.[6] By type of polyethylene, LDPE is 26% of the market, LLDPE is 32% and HDPE is 42%.

Since ethylene and, hence, polyethylene is derived from oil even if oil usage as a fuel declines it will still be needed as a feedstock for plastics such as polyethylene. One possible solution is to make biopolythene, that is, polythene derived from natural plant feedstock.

While polythene is an invaluable material, its resistance to degradation is a problem. When buried in a landfill, polythene, like most polymers or plastics, it will last for thousands of years. Plastics end up everywhere in the environment. The garbage patch of plastics in the Pacific Ocean is well known[7] due to the eight million tons of plastic garbage that end up in the ocean each year [3]. Perhaps, less well known is that microplastic particles end up incorporated into the Arctic Sea ice, of which polyethylene comprises as much as all the other plastics combined [4]. Thus, the sorting and recycling of plastic must become increasingly important in future unless we plan to fill up even the farthest corners of the planet with plastics. Unfortunately of the 8.3 billion tonnes of plastics produced by 2016 only 9% has been recycled, while 12% has been incinerated, and the rest was simply discarded [5]. Maybe there is another option. It has recently been reported that caterpillars of the wax moth Galleria mellonella, which normally feed on beeswax, have a prodigious appetite for polyethylene, offering the possibility that moths can help us eat out our way of this problem [6].

---

[5] https://www.plasteurope.com/news/POLYETHYLENE_t229725/

[6] technavio insights, Global Polyethylene Market 2014-2018.pdf

[7] https://www.nationalgeographic.org/encyclopedia/great-pacific-garbage-patch/

References

1. Strong, A. B. (2000). *Plastics: Materials and processing*. Upper Saddle River: Prentice Hall ISBN 0-13-021626-7.
2. van Dulken, S. (2000). *Inventing the 20th century: 100 inventions that shaped the world. From the airplane to the zipper*. New York: New York University Press ISBN: 0-8147-8808-4.
3. Zuckerman, C. *Garbage swell*, National Geographic, April 2017.
4. Obbard, R. W., Sadri, S., Wong, Y. Q., Khitun, A. A., Baker, I., & Thompson, R. C. (2014). Global warming releases microplastic legacy frozen in Arctic Sea ice. *Earth's Future, 2*, 315–320. https://doi.org/10.1002/2014EF000240.
5. Geyer, R., Jambeck, J. R., & Law, K. L. (2017). Production, use, and fate of all plastics ever made. *Science Advances, 3*, e1700782 5 pages.
6. Bombelli, P., Howe, C. J., & Bertocchini, F. (2017). Polyethylene bio-degradation by caterpillars of the wax moth galleria mellonella. *Current Biology, 27*, R283–R293.

Chapter 32
Polypropylene

The last of the common thermoplastic polymers to be invented, polypropylene or propylene (PP) has a repeat unit similar to that of polyethylene except that a methyl (CH₃) group replaces one of the hydrogen atoms, see Fig. 32.1. This asymmetric monomer raises the possibility of producing polymers with different tacticity that is with different spatial arrangements of the methyl group, see Fig. 32.2. The methyl groups can be in repeating arrangement where all the methyl groups are either on only one side of the polymer chain, an arrangement referred to as *isotactic*, or alternate between the opposite sides of the polymer chain, an arrangement referred to as *syndiotactic*. In the latter case, one can draw the repeating unit or "mer" of the polymer chain that is twice as large as the mer shown in Fig. 32.1, see Fig. 32.3. The final arrangement in Fig. 32.3 is called *atatic,* which is a random arrangement of the methyl groups.

These different tacticities affect the physical and mechanical properties. Fully isotactic polypropylene has a melting point of 171 °C, whereas syndiotactic polypropylene, which typically exhibits 30% crystallinity, has a melting point of 130 °C. Commercial polypropylene, which is largely isotactic and has around 50–60% crystallinity, has a melting point that depends on the amount of atactic material and the exact crystallinity, and ranges from 160–166 °C.

Researchers were working on making polypropylene in both the U.S.A. and Europe in the early 1950s. American chemists John Paul Hogan (1919–2012) [1] and Robert L. Banks (1921–1989)[1] working at the American company Phillips Petroleum appear to have first polymerized propylene. They filed a patent in 1951, which was only granted 32 years later in 1983! [2].[2,3] In 1953, the Italian chemist Giulio Natta (1903–1979),[4] building on work by the German chemist Karl Ziegler (1898–1973)

---

[1] https://www.acs.org/content/acs/en/education/whatischemistry/landmarks/polypropylene.html

[2] http://pubs.acs.org/doi/abs/10.1021/cen-v061n012.p007b

[3] discovery-of-polypropylene-and-development-of-high-density-polyethylene-commemorative-booklet.pdf

[4] https://www.nobelprize.org/nobel_prizes/chemistry/laureates/1963/natta-bio.html

© Springer International Publishing AG, part of Springer Nature 2018
I. Baker, *Fifty Materials That Make the World*,
https://doi.org/10.1007/978-3-319-78766-4_32

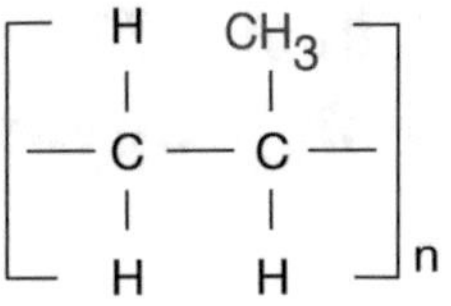

Fig. 32.1 The polypropylene monomer

$$\left[\begin{array}{ccc} & H & CH_3 \\ & | & | \\ - & C - & C - \\ & | & | \\ & H & H \end{array}\right]_n$$

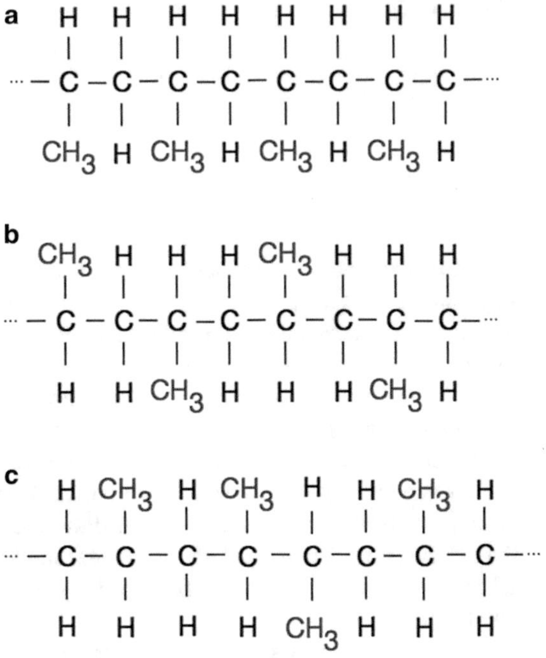

Fig. 32.2 Three different spatial arrangements of the methyl group in the polymer polypropylene: (**a**) isotactic, (**b**) syndiotactic, and (**c**) atactic or random

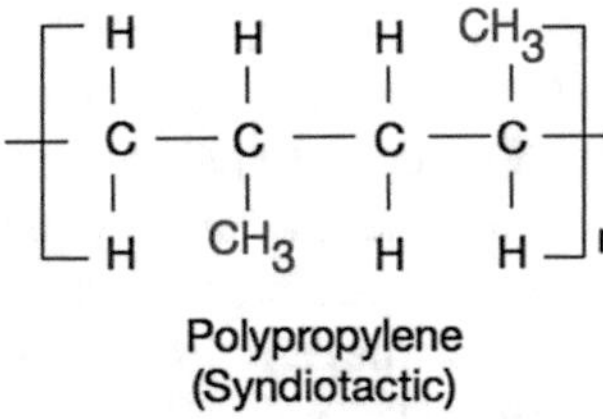

a

$$\cdots - \underset{CH_3}{\overset{H}{C}} - \underset{H}{\overset{H}{C}} - \underset{CH_3}{\overset{H}{C}} - \underset{H}{\overset{H}{C}} - \underset{CH_3}{\overset{H}{C}} - \underset{H}{\overset{H}{C}} - \underset{CH_3}{\overset{H}{C}} - \underset{H}{\overset{H}{C}} - \cdots$$

b

$$\cdots - \underset{H}{\overset{CH_3}{C}} - \underset{H}{\overset{H}{C}} - \underset{CH_3}{\overset{H}{C}} - \underset{H}{\overset{H}{C}} - \underset{H}{\overset{CH_3}{C}} - \underset{H}{\overset{H}{C}} - \underset{CH_3}{\overset{H}{C}} - \underset{H}{\overset{H}{C}} - \cdots$$

c

$$\cdots - \underset{H}{\overset{H}{C}} - \underset{H}{\overset{CH_3}{C}} - \underset{H}{\overset{H}{C}} - \underset{H}{\overset{CH_3}{C}} - \underset{CH_3}{\overset{H}{C}} - \underset{H}{\overset{H}{C}} - \underset{H}{\overset{CH_3}{C}} - \underset{H}{\overset{H}{C}} - \cdots$$

Fig. 32.3 The polypropylene monomer for the syndiotactic polymer

$$\left[\begin{array}{cccc} H & H & H & CH_3 \\ | & | & | & | \\ C - & C - & C - & C \\ | & | & | & | \\ H & CH_3 & H & H \end{array}\right]_n$$

Polypropylene
(Syndiotactic)

on organometallic catalysts was able produce isotactic polypropylene. The latter two later shared the 1963 Nobel Prize in Chemistry for their work on polymers. Later, Professor Natta also produced syndiotactic polypropylene. So called Ziegler–Natta catalysts are currently used to produce polypropylene.[5] In 1957, the Italian company Montecatini, now Montedison company, who sponsored Natta's work, were the first to produce isotactic polypropylene commercially in a plant in Ferrara, Italy - Phillips Petroleum did not start commercial production until the 1960s.[6]

---

[5] https://www.britannica.com/science/Ziegler-Natta-catalyst

[6] https://www.britannica.com/biography/Karl-Ziegler

Polypropylene is bit stronger at 25–40 MPa and has a higher elastic modulus (the resistance to stretching) at 0.9–1.5 GPa than its main rival low-density polyethylene, which has a strength and elastic modulus of 5–25 MPa and 0.1–0.3 GPa, respectively. Polypropylene is lightweight with a density of 930–950 kgm^{-3}, and so has a good strength-to-weight ratio for a polymer. It exhibits excellent elongation at room temperature of 150–300% with a reasonable impact resistance. Polypropylene also has good resistance to fatigue, and, thus, is often used in hinges such as those employed on flip-top bottles for drinks, detergents and other uses. Polypropylene also has a high resistance to a number of chemicals, including acids, organic solvents and electrolytes. It has a glass transition temperature (the point below which a rubbery polymer turns into a brittle glassy one) around the freezing point of water and so becomes brittle below that temperature. This is much higher than the −100 °C at which polyethylene becomes glassy and brittle.

The mechanical properties of polypropylene can be improved by the addition of inexpensive ingredients. For instance, an addition of 10% calcium carbonate nanoparticles to polypropylene can increase the elastic modulus by nearly 40% and reduce the brittle-to-ductile transition temperature [3]. Conversely, polypropylene fibers can be used to both increase the strength and reduce both cracking and spalling of concrete.

A foam form of polypropylene called expanded polypropylene (EPP) is available, which has a very low density in the range 20–120 kgm^{-3} and a very low elastic modulus of 75–500 MPa. It has excellent impact characteristics[7] such that EPP readily resumes its shape after an impact. It is, thus, used in many model aircraft and other radio-controlled vehicles where it can absorb the impact of crashes with little damage.

Over 50% of polypropylene is used in packaging, 11–12% is used in automotive applications, 10% in electronics/electrical applications, 12% in consumer products and 5% in construction. The uses are enormous and include ropes, thermal underwear, carpets, stationery, reusable containers, laboratory equipment and loudspeakers. Even banknotes are made from polypropylene in some countries with Australia leading the way by issuing banknotes for general circulation in 1992,[8] see Fig. 32.4. The switch to a polymer was not only because of the increased durability compared to paper but also to improve the security against counterfeiting. Because polypropylene doesn't melt below 160 °C it can used to make items such as like dishwasher-safe food containers and laboratory items that have to autoclaved – in contrast, polyethylene will soften at around 100 °C.[9,10]

Polypropylene is the second most produced polymer after polyethylene accounting for 23% of the 269 million tonnes World plastic production in 2015.[11] In 2013

---

[7] Optimization of Expanded Polypropylene Foam Coring to Improve Bum.pdf

[8] http://banknotes.rba.gov.au/australias-banknotes/history/

[9] Polypropylene- Is it different from Polyethylene?.pdf https://www.globalplasticsheeting.com/our-blog-resource-library/bid/92169/polypropylene-is-it-different-from-polyethylene

[10] Everything You Need To Know About Polypropylene (PP) Plastic.pdf, https://www.creativemechanisms.com/blog/all-about-polypropylene-pp-plastic

[11] The Plastic Industry Berlin Aug 2016 - Copy.pdf

Fig. 32.4 Australian polypropylene banknotes, which were introduced in 1992, were the first polymer banknotes in general circulation

61 million tonnes of polypropylene were produced with production expected to rise 79 million tonnes by 2018. This is expected to produce a rise in revenue over this period from $79 billion to $120 billion, a compound annual growth rate in revenue of 8.9%.[12] In 2015 China accounted for 50% of the demand for polypropylene and produces 28% of all plastics. The significantly, higher growth rate in the use of polypropylene compared to polyethylene suggests that at some point in the not too distant future polypropylene might be the most produced polymer. Apart from polypropylene's superior properties, this more rapid increase in polypropylene's use is also because it is cheaper: in May, 2017 the price of polypropylene was $1028 per tonne[13] compared to $1338 per tonne for low density polyethylene.[14] The production of oil and gas by fracking in the U.S.A. has led to a surge in polypropylene production and to current Global overcapacity and low prices.[15]

Polypropylene has a resin symbol of "5" and is recyclable, the recycled material being mixed with virgin material for further use.[16] It is always a challenge to separate different polymers since they are largely carbon and hydrogen sometimes with some nitrogen or oxygen. One way to separate PET bottles from polypropylene bottles is on the basis of density: PET has a density of 1430–1450 kgm^{-3}, whereas polypropylene is only 930–950 kgm^{-3} and, thus, floats on water (density = 1000 kgm^{-3}). If not recycled polypropylene will undergo degradation of the polymer chain if exposed to heat and ultraviolet radiation such as from sunlight. Fortunately, it is non-toxic.

---

[12] technavio insights, Global Polypropylene Market 2014-2018.pdf

[13] https://www.platts.com/news-feature/2014/petrochemicals/pgpi/polypropylene

[14] https://www.platts.com/news-feature/2014/petrochemicals/pgpi/ldpe

[15] http://news.ihsmarkit.com/press-release/chris-geisler/global-oversupply-polyethylene-polypropylene-challenging-margins-produce

[16] https://www.azocleantech.com/article.aspx?ArticleID=240

References

1. Stinson, S. (1987). Discoverers of polypropylene share prize. *Chemical and Engineering News, 65*(10), 30. https://doi.org/10.1021/cen-v065n010.p030.
2. US Patent 2825721, *Polymers and production thereof.* https://www.google.com/patents/US2825721
3. Eirasa, D., & Pessan, L. A. (2009). Mechanical properties of polypropylene/calcium carbonate Nanocomposites. *Materials Research, 12*, 517–522.

Chapter 33
Polystyrene

You can't easily tell one plastic from another simply by looking at it, but most of us would easily recognize polystyrene in its foam form called expanded polystyrene or by its (Dow Chemical Company) trademarked name Styrofoam, see Fig. 33.1. Polystyrene, as its name suggests, is made from the monomer styrene, see Fig. 33.2. Styrene, a sweet smelling oily liquid, was first produced by M. Bonastre in 1831 via distillation of storax or styrax balsam, which is the resin of the *Liquidambar* genus of trees.[1] Styrene, whose systematic name is Ethenylbenzene, is now one of most widely manufactured chemicals at around a 35 million tonnes annual production and growing at around 5% per year.[2] It is now usually made by dehydrogenation of ethylbenzene. Ethylbenzene occurs naturally in petroleum, but most is produced by combining benzene and ethylene.

The thermoplastic polymer polystyrene was accidentally discovered in 1839 by a German pharmacist Eduard Simon (1789–1856). Having produced storax by distillation, several days later he found that it had thickened into a jelly. He presumed that this thickening had occurred by oxidation and so he called the result "Styroloxyd" (styrol oxide). Soon afterwards in 1845, the German chemist August Wilhelm von Hofmann (1818–1892) [3,4]and his English student John Buddle Blyth (1814–1871) showed that "Styroloxyd" had the formula C_8H_8[5] and that its formation did not require oxygen [1].[6] Later, in 1866 the French chemist and politician Pierre Eugène Marcellin Berthelot (1827–1907) showed that the formation of Styroloxyd from styrol was by polymerization, leading to the name polystyrene.[7] By the 1930s, the

---

[1] Polystyrene – The Plastics Historical Society.pdf

[2] technavio insights, Global Styrene Market 2014-2018.pdf

[3] http://www.encyclopedia.com/people/science-and-technology/chemistry-biographies/august-wilhelm-von-hofmann

[4] https://www.britannica.com/biography/August-Wilhelm-von-Hofmann

[5] http://www.plasticseurope.org/what-is-plastic/types-of-plastics-11148/polystyrene/history.aspx

[6] http://extrudedpolystyrene.com.au/polystyrene/

[7] PlasticsEurope - Polystyrene (PS) - PlasticsEurope.pdf

© Springer International Publishing AG, part of Springer Nature 2018

I. Baker, *Fifty Materials That Make the World*,

https://doi.org/10.1007/978-3-319-78766-4_33

Fig. 33.1 Expanded polystyrene in different shapes

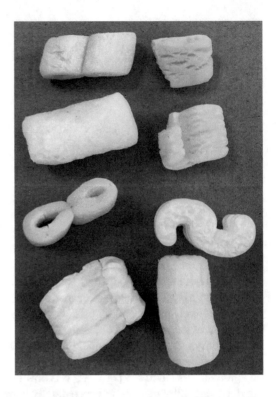

Fig. 33.2 The Styrene monomer from which polystyrene is made. The hexagonal ring representing the benzene molecule has a carbon atom at each corner, which is joined to the two adjacent carbon atoms and one hydrogen atom

German company IG Farben and the American company Dow Chemicals were producing styrene commercially from benzene and ethylene, a process first devised by Berthelot in 1851.[8] Production of polystyrene, a light, stiff thermoplastic, followed soon after. The process to make foamed polystyrene or expanded polystyrene, Styrofoam, was invented in 1941 by Dow Chemical. A hydrocarbon, typically pentane, is added to polystyrene beads as a foaming agent. On contact with steam, this turns into a gas and expands the beads by 40–50 times.[9]

Styrene will polymerize upon heating in an oxygen-free atmosphere, but commercially catalysts are used to make sure the process proceeds to completion. The polymerization is by an addition reaction, like polyethylene. In fact, the styrene

[8] styrene | chemical compound | Britannica.com.pdf

[9] http://epsa.org.au/about-eps/what-is-eps/how-is-eps-made/

monomer, is the same as the ethylene monomer of polyethylene but with a hydrogen replaced with a benzene group, see Fig. 33.2. However, the processing of the two seemingly similar polymers is quite different.

China accounts for around half the Global production of styrene. Around 60% of the styrene produced is made into polystyrene, which is widely used in injection-molded or foamed articles.[2] Most of the rest is copolymerized with other com-pounds—for example, with acrylonitrile and butadiene to produce acrylonitrile-butadiene-styrene (ABS) terpolymer, a hard, tough engineering plas-tic; with butadiene to make styrene-butadiene rubber, a tough synthetic rubber used in automobile tires; or with acrylonitrile or maleic anhydride to make styrene-acrylonitrile copolymer or styrene-maleic anhydride copolymer, which have improved heat resistance.[3]

Polystyrene is a transparent thermoplastic that can be easily colored. The poly-styrene chains are held together by van der Waals forces. These weak forces allow the polymer chains to easily slide past each other producing a flexible and elastic material. The benzene ring essentially prevents alignment of the molecules, that is, crystallization and so PVC is 100% amorphous [2]. (Non-expanded) polystyrene has good strength for a polymer of 36–52 MPa [3], and is significantly stronger than high density polyethylene (HDPE), the most used polymer, which has a strength of 30 MPa. For a polymer, it has a high elastic modulus (stiffness) of 3–3.5 GPa com-pared to only 1 GPa for HDPE. Unfortunately, it shows little ductility (1–2.5% elon-gation to failure) and has a low impact strength, as you may have noticed from your broken polystyrene CD cases. It can be cast into molds producing fine details very easily since it flows above 100 °C. Given these properties and that it is inexpensive, it is unsurprising that it is one of the most utilized polymers.

The applications for expanded polystyrene depend both on its shock absorbing properties and its light weight. Amongst other applications, it is used for coolers, as insulation for houses and other construction, for increasing the buoyancy of boats, for "take out" food containers, disposable plates and dishes, and packing "peanuts". However, you may have noticed more use of cardboard in "take out" food contain-ers, and inflated plastic bags being used instead of packing "peanuts". This is because of concerns about the very slow biodegradability of polystyrene even in sunlight, which has led to efforts to reduce or even ban its use.[10] There is no curbside recycling for polystyrene, and particularly as expanded polystyrene it is not eco-nomical to recycle because of its low density – you would be mostly recycling air unless it is compacted. The good news is that mealworms have been shown to digest Styrofoam and their excretions are safe for the environment [4].

While polystyrene's use in expanded form is very noticeable, it is actually used more in solid form for applications like CD cases (another item not long for this world), see Fig. 33.3, disposable cutlery and dinnerware, toys, computer cases and keyboards, and cell phones. Around 37% of polystyrene's use is in packaging, par-ticularly of food, while similar amounts of 11–13% are used each in domestic

---

[10] http://storyofstuff.org/blog/styrofoam-bans-are-sweeping-across-the-nation/

Fig. 33.3 A CD case, a common but rapidly declining use of polystyrene

appliances, consumer electronics and construction.[11] Around 13 million tonnes of (non-expanded) polystyrene will be produced in 2017 and an annual growth rate of 4% is expected (see footnote 2). By comparison, the market for expanded polystyrene will be around 8.5 million tonnes in 2017. Even though there are environmental concerns about expanded polystyrene, the market is expected to reach $19 billion by 2022, with a compound annual growth rate of over 6% from 2017 to 2022 (see footnote 2).[12]

References

1. Blyth, J., & Hofmann, A. W. (1843). On Styrole, and some of the products of its decomposition. *Memoirs and Proceedings of the Chemical Society (MPCS)., 2,* 334–358. https://doi.org/10.1039/MP8430200334.
2. Strong, A. B. (2000). *Plastics, materials and processing* (2nd ed.). Upper Saddle River: Prentice Hall ISBN: 0-13-021626-7.
3. Callister, W. D. (2001). *Fundamentals of materials science and engineering.* New York: Wiley ISBN: 0-471-39551-X.
4. Yang, Y., Yang, J., Wu, W.-M., Zhao, J., Song, Y., Gao, L., & Jiang, L. (2015). Biodegradation and mineralization of polystyrene by plastic-eating mealworms: Part 1. Chemical and physical characterization and isotopic tests. *Environmental Science and Technology, 49,* 12080–12086. https://doi.org/10.1021/acs.est.5b02661.

---

[11] http://www.essentialchemicalindustry.org/polymers/polyphenylethene.html

[12] http://www.marketwatch.com/story/expanded-polystyrene-market-worth-1897-billion-usd-by-2022-2017-04-27-72033113.

Chapter 34
Polytetraflouroethylene

You may never have heard of polytetraflouroethylene (PTFE), which is quite a mouthful, but you have surely used it when you use a non-stick frying pan or sauce-pan coated with it. You may have also used it as sealing tape in plumbing. PTFE was invented by accident in 1938 at the American company E. I. du Pont de Nemours and Company, Wilmington, DE, by Roy Plunkett (1910–1994) who was trying to develop a new refrigerant [1]. PTFE was patented in 1941 and in 1945 DuPont registered Teflon [2] as its trademark for this material.[1]

PTFE is a close cousin of polyethylene, in which all the hydrogen atoms in the polyethylene are replaced by fluorine atoms. The precursor of PTFE is tetraflouro-ethylene (TFE) a colorless, odorless, potentially explosive gas, see Fig. 34.1, in which the two carbon atoms have two bonds between them. A process called "free radical vinyl polymerization" is used to break the double bonds, see Fig. 34.1. The free bond then joins to other carbon atoms making a polymer chain, see Fig. 34.1. The fluorine atoms are very tightly held by the carbon atoms and PTFE chains are long and stiff resulting in a polymer in which the chains are so tightly packed together that it has the highest density of any plastic (2200 kg/m^3 for PTFE versus 910–960 kg/m^3 for polyethylene) [2]. The fluorine atoms in PTFE do not like to be near other atoms leading to it's non-stick nature. The carbon-fluorine bonds are very strong. This means that PTFE cannot burn (the carbon-oxygen bond is weaker) and is resistant to attack by corrosive chemicals [2]. Thus, it is frequently used for containers and hoses for reactive chemicals. Its non-reactivity and non-toxicity means that it is extensively utilized in the dairy, brewery and pharmaceutical industries. On the other hand, PTFE's inertness has the downside that PTFE lives in the environment forever, and it is more difficult to recycle than polyethylene.

PTFE has the interesting property that it has a very low coefficient of friction 0.05–0.1 (the third lowest of any material). One application that utilizes this low coefficient of friction is as a low maintenance ice for the skating surface in "ice" rinks, see Fig. 34.2. This property also makes it useful in many applications where

---

[1] http://www2.dupont.com/Products/en_RU/NonStick_Coatings_en.html

© Springer International Publishing AG, part of Springer Nature 2018
I. Baker, *Fifty Materials That Make the World*,
https://doi.org/10.1007/978-3-319-78766-4_34

Fig. 34.1 (**a**) The gaseous precursor tetraflouroethylene, and (**b**) the mer of the polymer polytetraflouroethylene

Fig. 34.2 An artificial ice hockey rick at Dartmouth College and a higher magnification photo showing the scratches on the surface from skate blades. The rink is made of sheets of PTFE that were donated by J. Asch. Courtesy R.L. Whitmore

it can be used to reduce the friction and wear such as bearings and gears. PTFE is an excellent insulator with an electrical resistivity of 1×10^{23} to 1×10^{25} Ohm.m, compared to 6.4×10^{2} Ohm.m for the semiconductor silicon and 2.4×10^{-8} Ohm.m for gold.[2] About 50% of PTFE's use is in aerospace and electronics applications, applications that utilize its excellent dielectric properties.

Teflon is not only a material, it has made its way into the vernacular after Pat Schroeder (1940-), a member of the U.S. House of Representatives from 1973–1997[3] referred to Ronald Reagan as the "Teflon President". The label has since been applied to many others.

Although PTFE is inert and non-toxic, the precursor from which it is produced Perfluorooctanoic acid (PFOA) (which is also a precursor to several other materials and chemicals) is both toxic and carcinogenic, and like PTFE lives indefinitely in the environment. Fortunately, some manufacturers have now discontinued the use of PFOA for making PTFE. Unfortunately, PFOA has spread around the world so that it is found on all continents and in creatures as diverse as polar bears in the Arctic and albatrosses in the middle if the Pacific Ocean. It is present at low levels in all Americans and its production has been the subject of a number of class action lawsuits.[4]

References

1. van Dulken, S. (2000). *Inventing the 20th century: 100 inventions that shaped the world. From the airplane to the zipper.* New York, NY: New York University Press ISBN: 0-8147-8808-4.
2. Strong, A. B. (2000). *Plastics, materials and processing* (2nd ed.). Upper Saddle River: Prentice Hall ISBN: 0-13-021626-7.

---

[2] http://hyperphysics.phy-astr.gsu.edu/hbase/Tables/rstiv.html

[3] http://www.nytimes.com/1990/07/01/magazine/the-prime-of-pat-schroeder.html?pagewanted=all

[4] http://www.liquisearch.com/perfluorooctanoic_acid/legal_actions/industry_and_legal_actions

Chapter 35
PVC

Polyvinyl chloride (IUPAC name: poly(1-chloroethene)), commonly referred to by its abbreviation PVC, sometimes has a poor reputation for its use as a cheap imitation leather, but PVC is invaluable for many applications.

PVC, a thermoplastic (it can be melted and set repeatedly) polymer, has a repeat unit similar to that of polyethylene except that a chlorine atom replaces one of the hydrogen atoms, see Fig. 35.1. PVC is one of the earliest synthetic materials that man has produced.[1] It was "discovered" several times and several attempts were made to commercialize it before its eventual successful commercial introduction in the 1930s. In 1838, the French chemist and physicist Henri Victor Regnault (1810–1878) produced vinyl chloride, and appears to have accidentally produced PVC in the form of a white powder [1]. Later, in 1872 the German chemist Eugen Baumann (1846–1896) also accidentally produced a white powder that was likely PVC.[2] In 1913, the German Friedrich Heinrich August Klatte was the first to receive a patent for producing PVC using sunlight for the polymerization process. The American chemist Waldo Lonsbury Semon (1898–1999) while working at B.F. Goodrich was able to plasticize the usually brittle PVC, which led to B.F. Goodrich commercializing the polymer and the start of its use in numerous applications.[3]

PVC comes in two varieties called rigid (RPVC) or unplasticized (uPVC or PVC-U), and flexible PVC. Pure PVC, which is a white, brittle polymer, can be made flexible and, hence, softer by the addition of plasticizers, the most common of which are phthalates, products of the organic acid phthalic acid [2]. Rigid PVC has good strength and elastic modulus (the resistance to stretching) compared to many other polymers. With a tensile strength of 41–45 MPa [3], it is stronger than polythene (15–40 MPa), although ABS plastics (37–110 MPa) and polystyrene can be stronger (30–100 MPa). Its elastic modulus (resistance to stretching) is similar to

---

[1] http://www.pvc.org/en/p/history

[2] http://www.issx.org/page/EugenBaumann

[3] http://lemelson.mit.edu/resources/waldo-semon

© Springer International Publishing AG, part of Springer Nature 2018
I. Baker, *Fifty Materials That Make the World*,
https://doi.org/10.1007/978-3-319-78766-4_35

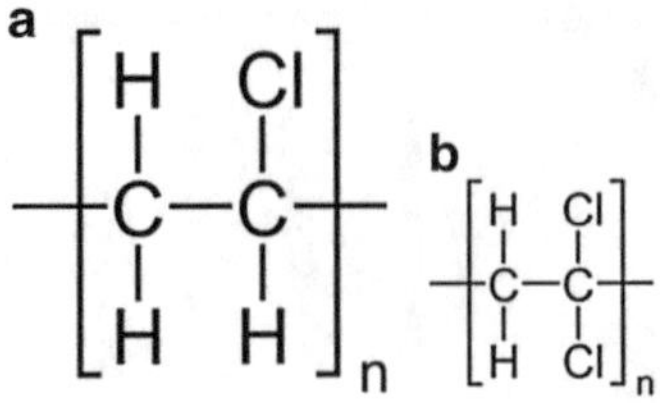

Fig. 35.1 (a) The polyvinyl chloride (PVC) and (b) the polyvinylidene chloride (PVDC) mers or repeat units. The mers are similar to polyethylene except PVC has one chlorine atom replacing a hydrogen and in PVDC two chlorine atoms replace two hydrogen atoms

polystyrene at 2.5–4.1 GPa, which is higher than that of other common polymers.[4] A key feature is that it has a higher fatigue strength (resistance to repeated loading) of 1.7 MPa for 10 million repetitions, compared to other polymers, but at the expense of a low impact strength of 2–8 KJ.m^{-2} at room temperature. However, the impact strength increases to 20 KJ.m^{-2} at -20 °C, making it useful for lower temperature use. It remains brittle up to its glass transition temperature (the temperature at which a material changes from brittle and glassy to rubbery) of approximately 87 °C. RPVC does not have good higher temperature properties and its use is limited to temperatures less than 60 °C.

Flexible PVC is significantly weaker than RPVC with a tensile strength of 6.9–25 MPa with a lower temperature resistance, but it has a higher impact strength, and it is easier to extrude or mold.[5] The properties of flexible PVC are more variable than RPVC because of the use by different manufacturers of different plasticizers, stabilizers (to enhance the poor heat stability), lubricants, fillers, pigments, and various processing aids.

The related polymer polyvinylidene chloride, which has two chlorine atoms in the mer rather than one as in PVC (see Fig. 35.1) was discovered accidentally in 1933 by Ralph Wiley while working at Dow Chemicals.[6,7] The best known application of polyvinylidene chloride was Saran Wrap, a food wrap, which was introduced in 1953. It was replaced by low density polyethylene in 2004 due to concerns about the decomposition of polyvinylidene chloride upon heating.

40–42% of the market for PVC is for pipes (particularly sewer pipes) and fittings because of its resistance to a wide range of chemicals including acids, bases, alcohols, salts and fats. Hosing and tubing accounts for another 20% of the market, rigid film and sheet for 16–19%, and (as flexible PVC) cable and wire insulation for 7–8%. Some consumer uses include bottles, non-food packaging, various bank or membership cards, imitation leather including in handbags, coats, shoes and car upholstery, flooring, music records, various inflatable products, and as a replacement for rubber in some applications.

---

[4] PVC Strength - PVC.pdf. http://www.pvc.org/en/p/pvc-strength

[5] PVC Properties - Vinidex.pdf. http://www.vinidex.com.au/technical/material-properties/pvc-properties/

[6] http://plastiquarian.com/?page_id=14257

[7] https://www.britannica.com/science/polyvinylidene-chloride

Fig. 35.2 A poncho made of PVC

   PVC, which is derived from petroleum, is the third most used polymer after polyethylene and polypropylene. It can be recycled roughly seven times, giving it a lifetime over 100 years. In 2013, 36 million tonnes of PVC were produced for revenue of $49 billion. Production is expected to rise to 45 million tonnes with a revenue of $62 billion by 2018,[8] representing a cumulative annual growth rate in revenue of 5.1%. The leading producers are China (39%), the U.S.A. (12%) and India (8%). In 2015, PVC accounted for 16% of the World production of plastics and 18.3% of the non-engineering plastics World market of 229 million tonnes.[9] China accounted for 28% of Global plastics production (Fig. 35.2).

---

[8] technavio insights, Global PVC Market 2014-2018.pdf
[9] The Plastic Industry Berlin Aug 2016 - Copy.pdf

References

1. Reif-Acherman, S. *The contributions of Henri Victor Regnault in the context of organic chemistry of the first half of the nineteenth century.* http://www.scielo.br/scielo.php?script=sci_arttext&pid=S0100-40422012000200037
2. Strong, A. B. (2000). *Plastics, materials and processing* (2nd ed.). Upper Saddle River: Prentice Hall ISBN: 0-13-021626-7.
3. Callister, W. D. (2001). *Fundamentals of materials science and engineering.* New York: Wiley ISBN: 0-471-39551-X.

Chapter 36
Rare Earth Magnets

The word magnet landed in English via the Latin *magnes,* which is derived from the Greek *magnēs lithos* meaning a stone (lodestone) from Magnesia, a Greek city.[1] Permanent magnets in the form of lodestone, a naturally-occurring weak magnet, were known to the Ancient Chinese, Romans, Greeks and, presumably, to other ancient societies. Lodestone consists mainly of ferrimagnetic magnetite (Fe_3O_4) with inclusions of ferromagnetic maghemite (γ-Fe_2O_3) that have been magnetized: interestingly, not all naturally-occurring magnetite has been magnetized. Although the magnetized lodestone could be shown to attract iron, Ancient Peoples made no practical use of these permanent magnets. The first practical use of a permanent magnet had to wait until eleventh century China (the Song dynasty) when a mariner's compass was developed that used a magnetized iron needle [1].

Apart from in the compass, the use of permanent magnets was very limited before the twentieth century partly because there just weren't any good permanent magnets. In 1931, a Japanese Metallurgist Tokushichi Mishima (1893–1975) produced an alloy containing (by weight) 58% Fe, 30% Ni and 12% Al that had a magnetic coercivity (a measure of the resistance of a ferromagnetic material to becoming demagnetized) of 400 Oe, which was nearly double that of the best magnetic steel of the time. He named this tough, durable and inexpensive material MKM steel, which is an acronym for "Mitsujima ka magnetic", Mitsujima ka being the name of the place where he grew up [2]. Mishima's invention led to the development of AlNiCo magnets in the 1940s. AlNiCo is somewhat of a misnomer since, while the magnets typically contain (by weight) 8–12% Al, 15–26% Ni and 5–24% Co, along with up to 6% Cu and up to 1% Ti, the largest constituent is iron.

AlNiCo magnets represent the first nanostructured engineered magnets. The Fe-Al-Ni-Co alloy system displays what is called a miscibility gap. At temperatures above 865 °C, AlNiCo alloys exist as a single phase body-centered cubic (b.c.c.) crystal structure, see Fig. 36.1. Below that temperature, the alloy splits into two b.c.c. phases that have slight differences in lattice parameter but large differences in

---

[1] Concise Oxford English Dictionary.

© Springer International Publishing AG, part of Springer Nature 2018
I. Baker, *Fifty Materials That Make the World*,
https://doi.org/10.1007/978-3-319-78766-4_36

Fig. 36.1 The body
centered cubic structure

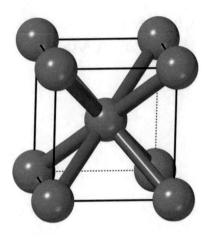

chemistry. One phase is Fe and Co rich and strongly ferromagnetic, while the other phase is rich in Ni and Al and only weakly magnetic. The decomposition occurs by a rather interesting process called spinodal decomposition. A common example of spinodal decomposition is a mixture of oil and water, which at higher temperatures form a single liquid, but which on cooling start to split into separate water and oil phases, which grow richer in oil and water with increasing time. In the case of the AlNiCo spinodal decomposition, small fluctuations in chemistry of the two phases increase with increasing time until two chemically different phases eventually form. The sizes of the phases do not grow significantly with increasing time. The result is two nanostructured phases that are aligned in specific directions in crystal, see Fig. 36.2.

AlNiCo magnets have an energy product, (BH_{max}), a measure of their strength (see Side Box) of up to 10 MGOe, but are fairly easily demagnetized. With the development of these high strength permanent AlNiCo magnets, it became possible to replace electromagnets with permanent magnets. This led to their widespread use in applications such as electric motors, loudspeakers, electric guitar pickups, microphones, sensors, and traveling wave tubes in microwave amplifiers.

A range of AlNiCo alloys have been developed over time. All are hard and brittle and, thus, are produced by either casting or sintering powders of the appropriate composition: the sintered products are finer grained and mechanically stronger but have inferior magnetic properties. Whatever processing route is used the magnets are subsequently given a three-stage heat treatment to develop the appropriate microstructure and magnetic properties [3]. While there are many AlNiCo alloys, which are often simply numbered AlNiCo 5, AlNiCo 9 etc., they can be broadly split into two categories, isotropic AlNiCos (whose properties are the same in all directions) that have a BH_{max} (see Side Box) of 1.25–3 MGOe and good resistance to demagnetization and anisotropic AlNiCos (whose properties vary with direction) that have a BH_{max} of 5–10 MGOe and very good resistance to demagnetization. The latter are produced by annealing in a magnetic field, which causes the Fe-Co phase to align its cube axes and become elongated along the applied field direction and at high fields form rods.

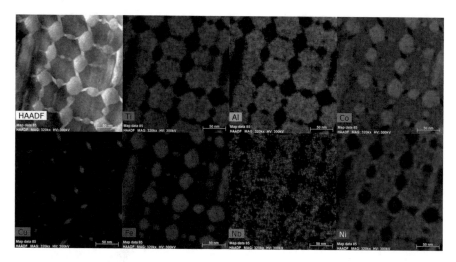

Fig. 36.2 High Angle Angular Dark Field (HAADF) image taken in a scanning transmission electron microscope showing the microstructure of an AlNiCo alloy. The colored images are X-ray fluorescence maps from the same region, which show the locations of the constituent elements. (Courtesy Lin Zhou and Matthew J. Kramer)

The heyday of AlNiCo alloys was relatively short lived. In the early 1950s hard ferrite or ceramic magnets were introduced. These magnetic hexagonal-structured iron-oxide compounds, mostly $BaFe_{12}O_{19}$ or $SrFe_{12}O_{19}$,[2] (see Fig. 36.3) exhibit lower BH_{max} values of ~2–8 MGOe than AlNiCo alloys, but they have some huge advantages: they are inexpensive, easy to process, chemically inert, have a strong resistance to demagnetization (high coercivity), and good high temperature properties because of their relatively high Curie temperature (the temperature at which a material loses its ferromagnetic behavior) of 577 °C. Hard ferrites, even though they are hard and brittle, rapidly took over the permanent magnet market from AlNiCo alloys and today dominate the permanent magnet market in terms of volume, accounting for 80% of all permanent magnets, although the far more powerful and expensive Rare Earth magnets have a higher market value.

AlNiCo alloys still find uses because of their large saturation magnetization of 2 Tesla, high resistance to demagnetization and because they are the only permanent magnets that can be used when red hot: AlNiCo magnets have the highest Curie temperature at 800 °C of any permanent magnet. Applications today are focused on loudspeakers, pickups, sensors, MRI machines, and very high temperature applications.

In 2015, the World market value of AlNiCo magnets sold was $158 million, this is expected to decrease to $121 million by 2020, an annual decrease of 5.3%. By comparison, the materials that ousted AlNiCo, hard ferrites, are expected to grow in sales from $5.8 billion in 2015 to $8.6 billion by 2020, and annualized growth rate of 8.3% [4].

---

[2] Magnetic-Materials-Background-1-History.pdf

Fig. 36.3 The complex
crystal structure of the hard
permanent magnet,
hexagonal-structured
$BaFe_{12}O_{19}$

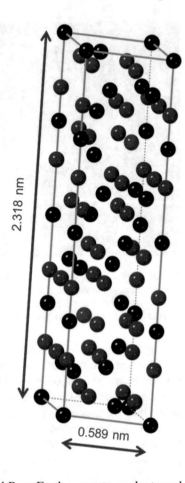

2.318 nm

0.589 nm

The development of the much more powerful Rare Earth magnets can be traced
to the work of Karl Josef Strnat (1929–1992) who discovered the magnetic compounds YCo_5 [2] and $SmCo_5$[3] while working at Wright Patterson Air Force Base in
Dayton, Ohio, in 1965 and 1966.[4] $SmCo_5$ has an energy product up to 25 MGOe and
a very high resistance to demagnetization. The invention of Sm-Co magnets led to
significant research efforts around the world [1]. Eventually in 1972, Dr. Karl Strnat,
then at the University of Dayton, along with Dr. Alden Ray developed Sm_2Co_{17}
magnets [5]. These magnets, which also contain small amounts of iron, copper,
zirconium and zirconium, have BH_{max} values as high as 33 MGOe and a high
resistance to demagnetization. Sm-Co magnets have excellent thermal stability,
good corrosion and oxidation resistance and can be used at temperatures as high as

---

[3] Magnet Energy/ PM history.pdf

[4] https://www.udri.udayton.edu/News/1999/Pages/DaytonContributestotheHistoryofMagnetic
Materials.aspx

Fig. 36.4 The hexagonal crystal structure of SmCo$_5$

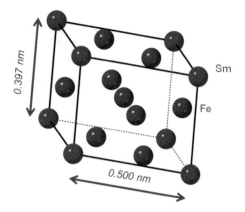

350 °C. However, because of their complex crystal structures, see Fig. 36.4, they are brittle with a low tensile strength of 35 MPa, and can crack and chip if subject to mechanical or thermal stresses. Thus, they are typically made by sintering powders rather than using a mechanical processing route. Recent developments include refining the grains in Sm$_2$Co$_{17}$ so that they are nanocrystalline. This produces a change from the rhombohedral crystal structure to hexagonal and significantly increases the energy product [6].

Sm$_2$Co$_{17}$ magnets had captured much of the market in high performance motors and high-temperature sensors when civil unrest in Zaire in 1978 disrupted the supply of cobalt producing a six-fold increase in the price of cobalt [7]. This accelerated the interest in developing non-cobalt containing permanent magnets and in 1982 neodymium iron boron magnets were invented simultaneously by two researchers: the Japanese scientist Dr. Masato Sagawa (1943-) at Sumitomo Special Metals and the American researcher Dr. John Croat of General Motors. Neodymium iron boron magnets were first marketed in 1984 by Crucible Magnetics of Elizabethtown, Kentucky. These magnets, which have the formula Nd$_2$Fe$_{14}$B, can have a BH$_{max}$ values up to 52 MGOe and generally have a higher resistance to demagnetization than Sm-Co magnets [4]. They are much cheaper than the Sm-Co magnets because they have a relatively lower fraction of the rare earth (Nd) element. However, they have a maximum operating temperature of 200 °C and are susceptible to corrosion. To combat the latter problem, they are often plated with nickel, or coated, see Fig. 36.5. Like the Sm-Co magnets, neodymium iron boron magnets, which have a complex tetragonal crystal structure, are brittle since they have no easy way to deform at room temperature. However, they exhibit a higher room temperature tensile strength of around 75 MPa compared to Sm-Co magnets. They are either made from powders that are pressed and sintered to form bulk material, or, more recently, from pulverized melt-spun ribbons that are mixed with a polymer matrix and compression molded or injection molded to shape. The latter material has a lower energy product due to the dilution effect of the polymer. They also typically contain 0.2–0.4 wt.%

Fig. 36.5 A neodymium iron boron magnet. Such magnets are typically coated, often with nickel, to prevent corrosion

aluminum, 0.5–1% niobium and 0.8–1.2% dysprosium.[5] They may also contain cobalt, copper or gadolinium. For applications where a particularly high coercivity is required these magnets can contain up to 6% dysprosium to replace the neodymium. Dysprosium also improves their corrosion resistance.

The name Rare Earth is a misnomer. The abundance in the Earth's Crust of the "Rare Earths" used in magnets, that is neodymium (ranked 28th), dysprosium (ranked 43rd) and samarium (ranked 38th) are 0.0033% 0.00062% and 0.0006%, respectively. By comparison, the "common" metals tin (ranked 49th), lead ranked (37th) and gold (ranked 72nd) have abundances of 0.00022% (ranked 72nd), 0.00099% and 3.1 × 10[−7]%, respectively. Thus, "Rare Earths" are not particularly rare compared to many other elements.

One of the downsides of Rare Earth magnets is that most Rare Earth metals are mined in China and most Rare Earth magnets are produced there – by 2011 China controlled 97% of the Rare Earth market.[6] The price of dysprosium, which hovered at less than $100 per kilogram through the early 2000s, began a steady rise in 2010 to around $150 per kg, and then in 2011 the price jumped to nearly $3500 per kg before falling thereafter. Neodymium showed a similar trend rising steadily from around $10/kg in 2001 but then spiking at over $450/kg in 2011. This prompted the U.S. Department of Energy's Advanced Research Project Agency-Energy (ARPA-E) to start the Rare Earth Alternatives in Critical Technologies (REACT) program, which spent $31.6 million on 14 projects to develop cost-effective alternatives to Rare Earth Magnets[7]. A small amount of that money went to my lab to develop the metastable ferromagnetic tetragonal compound MnAl. The exact reason for the price spike for the Rare Earth's may be the source of debate, but the problem was solved to some extent by the opening of mines in the U.S.A. and Australia.

The high energy density of Rare Earth magnets means that the magnets can be small, leading to a huge range of applications.[8,9] Applications of Neodymium Rare Earth Magnets range from computer hard drive magnets, microphones, headphones,

---

[5] How Neodymium Magnets are made.pdf

[6] http://fortune.com/2011/11/18/molycorps-1-billion-rare-earth-gamble/

[7] P. Campbell, "Performance and Cost Assessment for Motors Using Alternative Rare Earth Free Magnets", Dr.PeterCampbell.pdf.

[8] http://e-magnetsuk.com/neodymium_magnets/applications.aspx

[9] http://www.magnetsource.com/Solutions_Pages/RAREERTHapps.html

loudspeakers and electro-acoustic pick-ups, motors (e.g. washing machines, drills, food mixers, vacuum cleaners, hand dryers), generators (e.g. Wind turbines, Wave Power, Turbo Generators, etc.), sensors, MRI and NMR machines, magnetic bearings, magnetic levitation, and alternators. Applications of samarium cobalt Rare Earth magnets include, computer disc drives, sensors, linear actuators, satellite systems, motors, particularly (where temporary stability is vital), generators, and actuators.[10] The main difference between the uses of NdFeB and SmCo magnets is that the former are stronger magnets and cheaper, while the latter can operate at higher temperatures because of the higher Curie Temperature of SmCo magnets of 720–800 °C compared to 310–400 °C for NdFeB magnets, whose use is limited to under around 100 °C. While it is hard to compare prices because magnets may be used in different applications, a 12.5 mm by 6.25 mm by 3.125 mm block of a SmCo magnet costs $1.40, whereas the same magnet size made from NdFeB costs $0.30.[11]

The search goes on for better or cheaper permanent magnets. For example, Professor J.M.D. Coey (1945-) of Trinity College, Dublin, invented $Sm_2Fe_{17}N_2$ magnets in which the nitrogen resides interstitially in the lattice. These have a maximum energy product of ~40 MGOe [4]. Because the compound decomposes above 450 °C, it is not possible to sinter to full density and, thus, it is used as a bonded magnet, that is, powder particles of SmFeN are glued together. This limits the energy product to around 20 MGOe[3].

With the rapidly growing use of wind turbines and hybrid and electric vehicles, there is strong demand for Rare Earth magnets. In 2015, the market value for NdFeB magnets was $7 billion and that for Sm-Co magnets $427 million. NdFeB magnets are expected to see annual growth rate of 9% through 2020 to a market value of $10.7 billion, while Sm-Co magnets are expected to see an annual growth of 4.3% to a market value of $529 million [4].

Side Box

To discuss permanent magnets, we need to define the maximum energy product BH_{max}, which is the key parameter used to describe the strength of such a magnet. BH_{max} is defined by the area of the largest rectangle that can be inserted into the second quadrant of a plot of the measured magnetic flux density, B, (units of Tesla) versus the applied magnetic field, H, (in units of Oe or kA/m) for the magnet, as shown in Fig. 36.A BH_{max} is an energy density and commonly has units of kJ/m^3 or MGOe.

---

[10] http://www.electronenergy.com/products/materials/samarium-cobalt/
[11] http://www.magnet4less.com/index.php?cPath=1_5

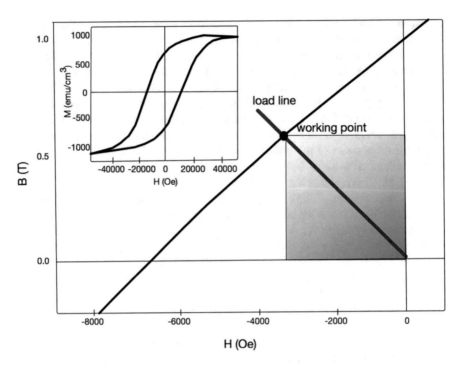

Fig. 36.A BH$_{max}$ is defined as the largest square that can be fitted into the second quadrant of the B-H loop (see inset). (After Jimenez-Villacorta and Lewis [3])

References

1. Coey, J. M. D. (2012). *Magnetism and magnetic materials.* Cambridge, England: Cambridge University Press. ISBN: 978-0-471-4771-9.
2. Cullity, B. D., & Graham, C. D. (2009). *Introduction to magnetic materials* (2nd ed.). Hoboken: Wiley. ISBN: 978-0-471-47741-9.
3. Jimenez-Villacorta, F., & Lewis, L. H. (2014). In J. A. Gonzalez Estevez (Ed.), *Chapter 7: Advanced permanent magnetic materials in nanomagnetism.* One Central Press (OCP) ISBN: (eBook): 9781910086056.
4. Kumar, A. (March 2016). *Permanent magnets: Technologies and global markets.* AVM029C. Wellesley, MA: BCC research report.
5. Livingston, J. D. (1990). The history of permanent-magnet materials. *Journal of Metals, 42,* 30–34.
6. Song, X. Y., Lu, N. D., Seyring, M., et al. (2009). Abnormal crystal structure stability of nano-crystalline Sm2Co17 permanent magnet. *Applied Physics Letters, 94*(2), 023102.
7. Constantinides, S. Permanent magnets in a changing world market. M*agnetics Magazine*, Feb 14, 2016.
8. Coey, J. M. D., & Sun, H. (1990). Improved magnetic properties by treatment of iron-based rare earth intermetallic compounds in ammonia. *Journal of Magnetism and Magnetic Materials, 87,* L251–L254.

Chapter 37
Rayon

Rayon was the first man-made fiber, see Fig. 37.1. However, it is not considered a synthetic fiber since it is made from regenerated cellulose, which is mostly derived from very high-cellulose wood pulp from softwoods. Although there are a number of steps to the manufacturing process, in essence the cellulose is converted to liquid form squeezed through small holes in a spinnerette and then converted back into cellulose fibers.

Rayon is often called artificial silk. The first patent for this artificial silk was granted in England in 1855 to the Swiss Georges Andemars.[1] The first commercially made rayon was produced in 1891 in Besançon, France. It was called "Chardonnet Silk" since it was developed by the French chemist Louis-Marie Hilaire Bernigaud, Comte de Chardonnet (1839–1924), who produced cellulose fibers by squeezing nitrocellulose, which is made by reacting cellulose with nitric acid through a spinnerette in 1889. Soon afterwards in 1890, the French chemist, Louis-Henri Despeissis, patented fibers made through processing cuprammonium rayon. These were commercialized in 1904 by the German company J.P. Bemberg A.G. as "Bemberg silk", the first artificial silk comparable to real silk.

Earlier, in 1891, three British chemists, Charles F. Cross (1855–1935), Edward J. Bevan (1856–1921), and Clayton Beadle (1868–1917), produced another variant of rayon, which is the type most commonly used today, that became known as "Viscose Rayon". This is made by dissolving cellulose xanthate in a dilute sodium hydroxide solution. This was first commercialized by the British company Samuel Courtauld & Company in 1905, and first produced in the U.S.A. in 1911 by the American Viscose Corporation. High performance rayon fibers were developed somewhat later in 1947.[2,3]

---

[1] https://www.thoughtco.com/history-of-fabrics-4072209
[2] Rayon | American Fiber Manufacturers Association.pdf
[3] https://www.britannica.com/topic/rayon-textile-fibre

© Springer International Publishing AG, part of Springer Nature 2018
I. Baker, *Fifty Materials That Make the World*,
https://doi.org/10.1007/978-3-319-78766-4_37

Fig. 37.1 Scanning electron micrographs of Rayon. Top: Bundles of Rayon Fibers; bottom: cross-sectioned Rayon fibers. (Courtesy of Mark P. Staiger)

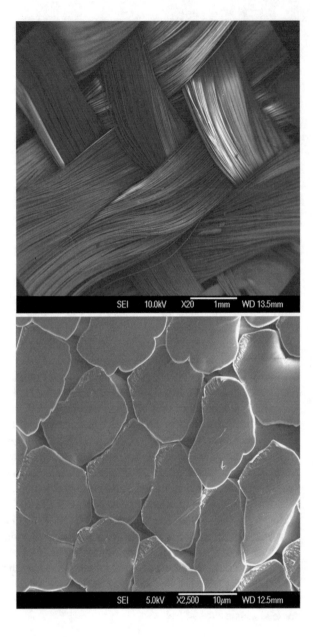

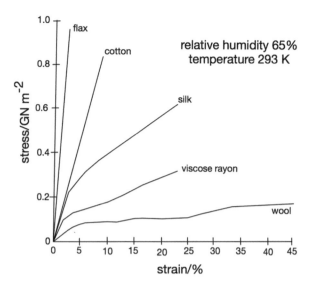

Fig. 37.2 Stress-strain curves for various single natural fibers. (After Weidmann et al. [1])

Rayon has properties somewhat similar to cotton, but it is cheaper to produce than wool, cotton or silk and requires fewer workers. Thus, in the 1920s, it contributed to the decline of the U.S. textile industry.[4]

Rayon has a lot of desirable properties for fabric. It is soft; easy to dye; drapes well; it is more moisture absorbent than cotton as long it is not coated with "wrinkle-resistant" chemicals, it is more biodegradable than cotton; and although it is "man-made" it is essentially a natural material.[5] Rayon fibers are stronger than wool but weaker than other natural fibers such as silk, cotton and flax [1], see Fig. 37.2.

Rayon has a number of uses in home furnishings such as sheets, curtains, upholstery; a wide variety of apparel, surgical products, tire cord and is sometimes blended with wood pulp to make paper. In 2014, textiles and apparel account for 45% of the use of rayon fibers, home furnishings account for 25%, industrial use was 15% and medical use was 15%. The rayon market in 2014 was worth $10 billion with a volume of 4.9 million tons. Rayon use is expected to increase at 9–10% to around $16 billion and 7.2 million tons by 2019.[6] China is the leading producer of rayon at 38% of the market, followed by India at 16%.

References

1. Weidmann, G., Lewis, P., & Reid, N. (Eds.). (1990). *Structural materials; materials in action series*. London: Butterworths The Open University, ISBN: 0-408-04658-9.

---

[4] http://study.com/academy/lesson/1920s-textiles.html

[5] Battle of the fabrics — rayon vs. cotton | The Chronicle Herald.pdf

[6] 2015–2019 Global Rayon Fibers Market, technavio insights.

Chapter 38
Rubber

As anyone who has stretched an elastic or rubber band knows the behavior of rubber is quite different to that of most other materials. If you pull on a crystalline material such as steel it can be stretched only about 0.1–0.2% before it remains permanently stretched: pull it less than 0.2% and the material will spring back to its original length (although this is such a small a change that you would be hard pressed to notice it). Rubber, sometimes called India Rubber, behaves very differently. It can be stretched over 500% and still behave elastically, that is it will spring back to its original length. What is happening when you do this? Rubber is a polymer chain that consists of repeat units termed mers of *cis*-1, 4 isoprene joined end-on-end in long chains, as shown in Fig. 38.1. (*Cis* means that the side chains are the same side of the molecule.) In its unstretched state these polymer chains are randomly oriented, all tangled up and can be said to have a high degree of disorder, as indicated schematically in Fig. 38.2a. When you pull on a piece of rubber you are stretching these polymer chains out so that they line up along the direction of stretching, as shown in Fig. 38.2b. The rubber molecules have also become more ordered. If you release the piece of rubber it will spring back to its original length. This process may seem fast but it isn't instantaneous. The polymer chains require thermal agitation, which allows the bonds to rotate and again become randomly oriented. The driving force for this process is that the molecules prefer to be disordered since this increases the entropy or randomness of the system. Commercially-used rubber typically has polymer chains cross-linked by vulcanization (see later). The cross-linking increases the strength as shown in Fig. 38.3 and provides anchor points between adjacent molecules so that they don't pull apart until at high stress. Even so, the elastic or Young's modulus, which is the resistance to being stretched, is very low compared to a metal, ceramic or semiconductor – 0.01–0.1 GPa versus 200 GPa for steel – and is even low compared to the value for many thermoplastic polymers of 1–4 GPa.

Rubber behaves this way because it is above its so-called 'glass transition temperature'. What happens when it is below this temperature? If you pull on rubber at $-200\,°C$, there is little thermal agitation and the polymer chains are frozen in place

© Springer International Publishing AG, part of Springer Nature 2018
I. Baker, *Fifty Materials That Make the World*,
https://doi.org/10.1007/978-3-319-78766-4_38

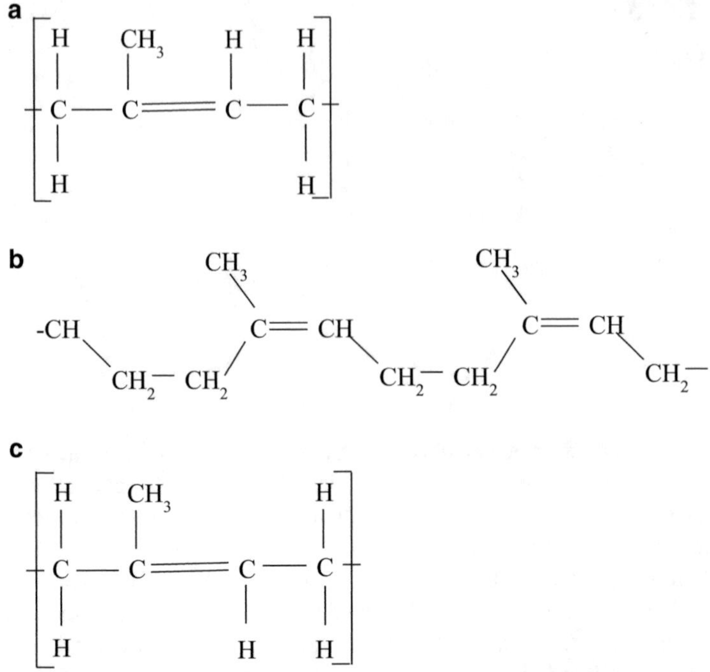

a

$$\begin{bmatrix} \overset{\displaystyle H}{\underset{\displaystyle H}{\vert}} & \overset{\displaystyle CH_3}{\underset{\displaystyle \vert}{\vert}} & \overset{\displaystyle H}{\vert} & \overset{\displaystyle H}{\underset{\displaystyle H}{\vert}} \\ C & - C = C - C \end{bmatrix}$$

b

$$\begin{bmatrix} \overset{\displaystyle H}{\underset{\displaystyle H}{\vert}} & \overset{\displaystyle CH_3}{\underset{\displaystyle \vert}{\vert}} & & \overset{\displaystyle H}{\underset{\displaystyle H}{\vert}} \\ C & - C = C - C \\ & & H \end{bmatrix}$$

c

Fig. 38.1 (**a**) The repeat unit cis-1, 4 isoprene that is the basis of (**b**) the natural rubber polymer chain, and (**c**) the repeat unit trans-1, 4 isoprene that is the basis of synthetic rubber

a

b

F ⟵   ⟶ F

Fig. 38.2 Schematic of polymer chains of rubber in (**a**) the unstressed state, and (**b**) when a force F is applied so that the chains attempt to line up with the force

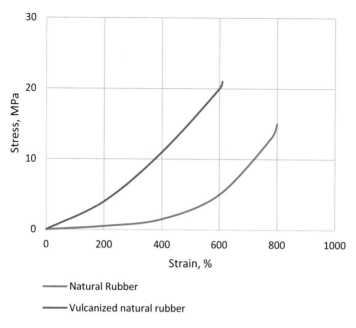

Fig. 38.3 Stress-strain curves for natural and vulcanized rubber

since the bonds between atoms cannot easily rotate. If you pull on the rubber band now, it is much more difficult to stretch and when you stretch the band only a small amount it will break.

An interesting twist on the effect of the glass transition temperature is to stretch a rubber band at room temperature and then fix the ends. If you now cool the rubber band to around −200 °C (−200 °C is a convenient temperature to use since this is the temperature of liquid nitrogen, which is commonly used for cooling since at $0.40 per litre it is cheaper than bottled water or milk) and then cut it, it does not spring back as it would at room temperature but the two halves remain fixed – the bonds are frozen in place. If we allow the rubber to warm to room temperature it will slowly go back to its original length as thermal agitation allows the bonds to reorient, as shown in Fig. 38.4.

Another unusual feature of rubber is its response to being heated. When most crystalline materials are heated they expand due to the increasing amplitude of the atomic vibrations. In contrast, rubber, like most elastomers, shrinks when heated. Heating again allows greater bond mobility, which allows the polymer chains to coil up more.

Although rubber had been used in Brazil for over 3600 years [1], the boom in the industrial use of rubber did not start to take off until the nineteenth century due to two developments: the masticator, a machine used to shred rubber invented by British Engineer Thomas Hancock (1786–1865), and vulcanization, a process for cross-linking rubber polymer chains using sulfur (see Fig. 38.5) invented almost

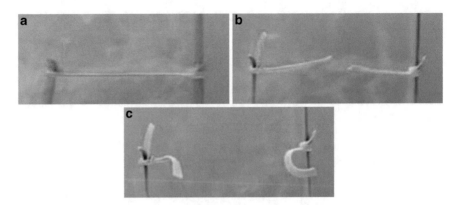

Fig. 38.4 (**a**) Rubber band stretched, attached to two rods and cooled to −200 °C; (**b**) immediately after cutting the band into two at −200 °C it remains quite rigidly in place; and (**c**) after allowing the rubber to warm to room temperature the two pieces of the rubber band contract

Fig. 38.5 Rubber polymer chains (**a**) before and (**b**) after cross-linking by sulfur atoms (Vulcanization)

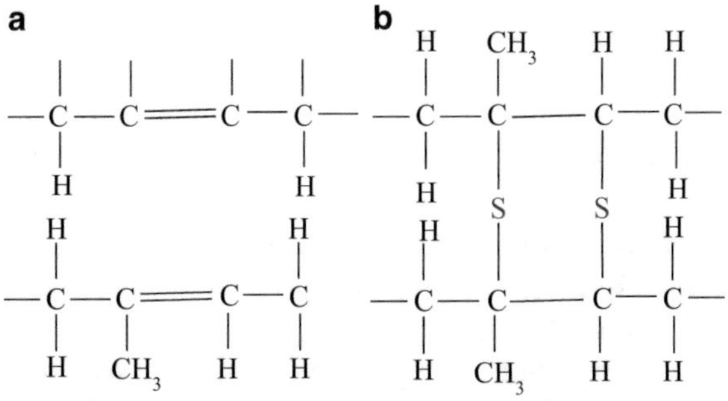

simultaneously by Hancock and an American chemist Charles Goodyear (1800–1880) [2].[1] Goodyear's patent was filed in the U.S.A. only 8 weeks after Hancock's British patent application.[2] The development of vulcanized rubber, which typically contains about three weight percent sulfur and is both stronger, more elastic and less sticky than natural rubber, proved useful for seals and later for tires. Tires, which currently account for well over 50% of the use of rubber, also incorporate carbon black for further strengthening. The carbon also makes them black. The initial rubber boom in Brazil was overtaken by production in Asia, a process initiated by Sir Henry Alexander Wickham (1846–1928): Wickham took 70,000 rubber tree seeds to the Royal Botanical Gardens at Kew, London, from where seedlings were dispatched to various tropical destinations in what was then the British Empire.[3,4]

[1] https://www.britannica.com/biography/Charles-Goodyear

[2] https://www.britannica.com/biography/Thomas-Hancock

[3] http://rainforests.mongabay.com/10rubber.htm

[4] http://ipkitten.blogspot.co.nz/2015/10/henry-wickham-amazon-river-and-rubber.html

Natural rubber is derived from latex, which is a secretion from the bark of a tree: Hevea brasiliensis. As the name suggests, the rubber tree is from Brazil, but most natural rubber is now derived from trees in South and South-East Asia, particularly Thailand and Indonesia, which account for more than half the total production.[5] China is the largest consumer utilizing ~33% of world production.[6] Natural rubber production is a very labor-intensive industry. However, a little over half of the rubber used is synthetic rubber (12.3 million tonnes of natural rubber and 14.4 million tonnes of synthetic rubber in 2015),[7] which is based on polymerization of the trans (trans means that the side chains are on the opposite sides) – 1,4 isoprene molecule (Fig. 38.1c). This is very capital intensive and employs few people compared to natural rubber production. There are other synthetic rubbers such as styrene-butadiene rubber, nitrile rubber and polychloroprene (neoprene) that have different side chains on the molecule, which lead to different mechanical properties and chemical behavior. The production of natural rubber has been falling a little each year for the last few years. The price of the synthetic rubber strongly depends on the price of butadiene,[8] which is derived from oil. Thus, the low price of oil for the forseeable future should favor the production of synthetic rubber.

References

1. Keoke, E. D., & Poterfield, K. M. (2002). *Encyclopedia of American Indians contributions to the world*. New York: Checkmark Books ISBN 0-8160-5367-7 (pbk).
2. Amato, I. (1998). *Stuff: The materials the world is made of*. New York: Avon Books ISBN-10: 0380731533.

---

[5] https://top5ofanything.com/list/7738a992/Rubber-Producing-Countries

[6] https://www.ihs.com/products/natural-rubber-chemical-economics-handbook.html

[7] WebSiteData_Feb2017New.pdf

[8] https://chemical-materials.elsevier.com/chemicals-industry-news-and-analysis/synthetic-rubber-producers-pressure-butadiene-prices-climb/

Chapter 39
Silicon

Until recently, silicon was a material that we all used in various electronic devices but rarely saw. However, the increasing deployment of photovoltaic solar panels means that we now often see silicon either in polycrystalline, monocrystalline or amorphous (non-crystalline) forms. In fact, the use of elemental silicon in solar cells has been growing so rapidly in recent years that more silicon is now used in solar cells than in electronic devices. Still, most silicon (80%) is not used in elemental form but as an alloying element in various metals principally iron (ferro-silicon) and aluminum (aluminum-silicon) and in making silicones, which are silicon-oxygen polymers. Silicon is also present as an oxide or silicate in ceramics, glasses and various building materials.

Silicon was originally named (before it was even isolated) silicium by Sir Humphrey Davy (1778–1829) in 1808 with the "ium" ending being used because it was originally expected to be a metal [1]. The British chemist Thomas Thompson (1773–1852) changed the name to silicon in 1817, because he thought that the element would not have the characteristics of a metal but that it would me more like carbon or boron, which are nearby in the periodic table [1]. Silicon was definitively isolated in amorphous form in 1824 by the Swedish chemist Jöns Jacob Berzelius (1779–1848) [1] by heating potassium fluorosilicate with potassium [2]. The gray metallic-looking crystalline form of silicon was not produced until 1854 by French Chemist Henri Sainte-Clair Deville (1818–1881)[1]. Silicon is now produced by reacting silica (SiO_2) with carbon at temperatures above 1900 °C, i.e. above the melting point of silicon of 1414 °C. Interestingly, silicon has the rare property that it is denser in liquid form than as a solid and, thus, like water, the solid floats on the liquid. Polycrystalline silicon can be made simply by casting. Single crystal boules of silicon up to 2 m long and up to 300 mm in diameter are produced from high-purity silicon using the Czochralski process in which a seed crystal of the required orientation contacts molten silicon and is slowly withdrawn from it, thus pulling out a crystal on the seed. The crystals are then sliced into wafers several hundred microns (a micron is one thousandth of a millimeter) in thickness.

---

[1] http://nautilus.fis.uc.pt/st2.5/scenes-e/elem/e01410.html

© Springer International Publishing AG, part of Springer Nature 2018
I. Baker, *Fifty Materials That Make the World*,
https://doi.org/10.1007/978-3-319-78766-4_39

Fig. 39.1 Temperature
dependence of the
electrical conductivity of
silicon

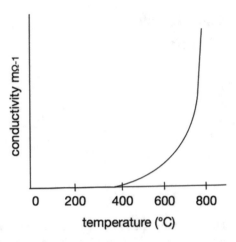

Silicon's thermal conductivity at 149 W/(m.°C) is similar to that of some metals such as magnesium (156 W/(m.°C) and higher than that of some other metals such as manganese at 7.8 W/(m.°C). However, the electrical conductivity of intrinsic silicon is much less at 1×10^3 Siemens/m than metals such as aluminum (3.8×10^7 Siemens/m) or iron (1×10^7 Siemens/m), but much greater than an insulator such as boron, which has an electrical conductivity of only 1×10^{-4} S.m^{-1}. Since the electrical conductivity of silicon is part way between that of an insulator and a conductor it is called a semiconductor. However, the electrical conductivity behavior of silicon is fundamentally different to that of metals. The conductivity of metals arises because of the large numbers of free electrons. For example, in copper there is one free electron per atom, producing 8.5×10^{28} electrons/m^3. Increasing temperature decreases the conductivity of copper as the free electrons traveling through the metal lattice interact with the increasingly vibrating atoms. In contrast, in an intrinsic semiconductor such as silicon the number of electrical carriers increases exponentially with increasing temperature and, thus, the conductivity shows the same trend, see Fig. 39.1.

Silicon adopts the diamond cubic structure, see Fig. 39.2, with each silicon atom bonded to four adjacent silicon atoms. Electrical conductivity in silicon (and other intrinsic semiconductors such as germanium) arises because thermal excitation frees one of the two shared electrons between neighboring Si atoms, thereby leaving a "hole" behind: increasing temperature frees more of these intrinsic electrons and holes. When a voltage is applied to an intrinsic semiconductor such as silicon, the holes and electrons flow (in opposite directions) and produce a small current. Unfortunately, the conductivity of intrinsic semiconductors is relatively low and, thus, they are not used for most electronic devices.

In extrinsic semiconductors, foreign atoms are added to produce extra electrons or holes that overwhelm the thermally-produced intrinsic electrons and holes, and, thus, control the conductivity at room temperature. Heavily-doped silicon, which is used in the source and drain regions in transistors, can have a conductivity of 1×10^6

Fig. 39.2 Crystal structure
of silicon, which is the
same as that of Diamond

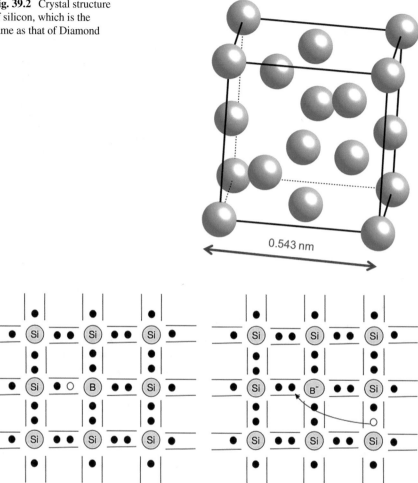

Fig. 39.3 In a p-type extrinsic semiconductor an atom such as boron is substituted into the silicon lattice. Since the boron only has three electrons (depicted as black dots), it is missing one electron compared to silicon. The result is a hole (depicted as an open circle). As indicated, electrons can migrate to fill this hole

Siemens/m. If we replace a silicon atom with a phosphorus atom, which has five valence electrons, the bonds to all adjacent silicon atoms are satisfied with one of the electrons from the phosphorus left over, which, thus, becomes an extrinsically-added electron. This type of material is called a *n*-type semiconductor since the extra electron carries a negative charge. The opposite type is the *p*-type, when we replace some silicon atoms with atoms such as boron, which have only three valence electrons. In this case, only three of the adjacent silicon atoms are bonded to and the fourth bond has a hole, which can migrate as a positive charge carrier, see Fig. 39.3.

Fig. 39.4 A silicon wafer

Most electronic devices are made of extrinsic semiconductors usually in combinations of *p* and *n* type semiconductors. The most well-known semiconductor device is the transistor. This was invented by William Schockley (1908–1989), John Bardeen (1908–1991) and Walter Brattain (1902–1987) in 1947 at Bells Labs. This first transistor was made not of silicon but from germanium and won its inventors the Nobel prize in Physics in 1956.[2] Although the first transistor was made of germanium most semiconductor devices are now made of silicon because it can withstand higher operating temperatures and higher power. Further, the silicon dioxide, which is grown on top of silicon is insoluble in water, and silicon is the second most common element after oxygen, comprising almost 28% of the Earth's crust with various silicates making up 90% of the Earth's crust [1].

The silicon used in semiconductor devices has impurity levels (before other atoms such as boron or phosphorus are added) of less than 1 part per billion. Whereas the silicon used in solar panels has impurity levels of 1 part per million [1]. Another difference, between solar panels and semiconductors devices is that all the silicon in semiconductors is monocrystalline, that is there are no grain boundaries or linear defects called dislocations, which both affect the electrical properties. The single crystals are grown in specific orientations either with the cube face or the body diagonal of the unit cell (see Fig. 39.2) along the growth direction. They are then sliced into wafers on which electronic devices are produced, see Fig. 39.4.

While single crystal silicon solar panels are more efficient (~25%) at converting light to electricity than the best polycrystalline solar panels (~20%) they are much more expensive so that currently only about 35% of the market is single crystal silicon solar panels. See Fig. 39.5. The latest generation solar panel may have on 2 micron thick layer of silicon. Thus, a one meter square solar panel contains only 2 mm³ of silicon. Even though silicon is strong in compression but very brittle, it is used in some mechanical watches and various microelectromechanical or MEMs devices.

---

[2] http://history-computer.com/ModernComputer/Basis/transistor.html

Fig. 39.5 Solar panel array in Vermont, U.S.A. where they are becoming as common in fields as cows. (Courtesy of Jennie Shurtleff)

A good case could be made that silicon is the most important material of the last few decades and that we are now living in the Silicon Age. There are certainly semi-conductors that are have higher electron mobility than silicon, and photovoltaic materials that are more efficient at conversion of light in electricity than silicon, but with the huge investment in factories in silicon processing, the vast expertise in manufacturing of this material and its low cost, growth in the use of silicon-based devices seems assured.

References

1. Emsley, J. (2001). *Silicon, nature's building blocks: An A–Z guide to the elements*. Oxford, England: Oxford University Press. ISBN: 0-19-850340-7.
2. Stwertka, A. (2012). *A guide to the elements*. Oxford, England: Oxford University Press. ISBN-10: 0199832528.

Chapter 40
Silver

Few probably remember "Silver Thursday", March 27th, 1980 when the collapse in the price of silver produced panic on the futures and commodity exchanges.[1] In 1979–1980, the American oil billionaire brothers Nelson Bunker Hunt (1926–2014)[2] and William Herbert Hunt (1929-) attempted to corner the silver market and it is estimated eventually held $10 billion worth of silver (195 million troy ounces).[3] This amounted to one third of the World silver supply, and led to a price increase from January 1st, 1979 of $6.08 per troy ounce to $49.94 per troy ounce a little over a year later on January 18th, 1980. The Hunt brothers borrowed heavily to purchase the silver and when the New York Commodity Exchange, Inc. (COMEX) placed restrictions on purchasing on margin, they were caught in a classic margin call debacle. This not only affected them but also could have led the collapse of several banks and brokerage companies on Wall Street that had lent to them. The government provided a $1.1 billion line of credit to the banks and brokerages to avert a disaster. Within 2 months of its high point, the price of silver had dropped to less than $11 per troy ounce. It did not end well for the brothers, who lost over $1 billion on their silver trading. The current price is $16.66 per troy ounce.

You can buy silver items in many places, but one of the most interesting places to purchase whatever you want made of silver is the subterranean London Silver Vaults, which are entered from an unassuming doorway on Chancery Lane in London.[4] The London Silver Vaults started life as the Chancery Lane Safe Deposit in 1885, a place to rent space for the storage of silver, jewelry and other items behind 1.2 m-thick steel-lined walls. The vaults soon transitioned to being the home of silver dealers and now house more than 40 shops, all of which have been family-owned for over 50 years, and likely house the world's largest silver hoard for sale.[5, 6]

---

[1] https://www.britannica.com/topic/Silver-Thursday

[2] https://www.washingtonpost.com/business/nelson-bunker-hunt-texas-oil-baron-who-lost-much-of-his-fortune-dies-at-88/2014/10/22/81739876-5a02-11e4-8264-deed989ae9a2_story.html

[3] http://www.wealthx.com/articles/2015/william-herbert-hunt/

[4] http://londonist.com/2015/01/have-you-ever-explored-londons-silver-vaults

[5] http://www.langfords.com/langfords/london-silver-vaults/

[6] https://londonist.com/2015/01/have-you-ever-explored-londons-silver-vaults

© Springer International Publishing AG, part of Springer Nature 2018
I. Baker, *Fifty Materials That Make the World*,
https://doi.org/10.1007/978-3-319-78766-4_40

Fig. 40.1 A $1 dollar silver certificate that at one time was redeemable for either silver coins or silver bullion

Silver has been used as coinage by many civilizations since around 600 B.C. when it was first used in Lydia,[7] an ancient kingdom that is now part of present day Turkey, as electrum, a naturally-occurring alloy of silver and gold. Silver coinage was often based on weight, a feature reflected in the British Pound and the Israeli Shekel (weights of 0.454 kg about 0.11 kg, respectively). It has also been used as a guarantee for bank notes. The U.S.A. issued silver certificates in various denominations from $1 to $1000 between 1878 and 1964, see Fig. 40.1, and for many years these were a major part of the currency in circulation. These could be redeemed for silver coins but for a single year from June 24th, 1967 to June 24th, 1968 they could be exchanged for silver bullion.[8] Silver is not currently used as currency.

Silver, symbol Ag from the Latin *Argentum*, [9] is the metal with the highest reflectively at wavelengths above 450 nm and has often been used for mirrors and optical coatings – currently, aluminum is used more because it has a high refelectivity over the whole visible spectrum and does not tarnish like silver. Silver also has the highest thermal conductivity (406 W/m.°C compared to 385 W/m.°C for copper and 50 W/m.°C for steel) and electrical conductivity (6.3 × 10^7 Siemes/m for silver compared to 6.0 × 10^7 Siemes/m for copper and 0.6 × 10^7 Siemes/m for steel) [10] of any metal.

Silver was first smelted 6–7000 years ago. It is 70 times more common in the Earth's crust than gold (0.08 parts per million (p.p.m.) versus 0.0011 p.p.m. for gold) but much less common than the other periodic table Group 11 element copper at 68 p.p.m. – Group 11 elements are sometimes referred to as the coinage metals. Although more common than gold, since it is more reactive than gold, it is much less common in elemental form and so in the Ancient world was more expensive

---

[7] http://www.britishmuseum.org/explore/themes/money/the_origins_of_coinage.aspx

[8] http://www.antiquemoney.com/silver-certificates/

[9] https://www.collinsdictionary.com/dictionary/english/argentum

[10] BBC Podcast Silver.

Fig. 40.2 A silver musical box. Notice the tarnish, which is typical of silver. (Courtesy Stephanie Turner)

than gold. Silver is found in high concentrations in the mineral argentite, which is silver sulfiide (Ag_2S) and is also present in nickel-copper ores. In the Ancient world, the largest source of silver was from lead ores, and at the beginning of the Roman Empire, one tonne of lead produced 4.5 kg of silver. Nowadays, silver is obtained from its ore either by chemical leaching or by smelting. It is also produced during the processing of both lead and copper.

Silver is ductile and malleable, but has low strength and so it is not very useful structurally. However it has other uses than in coinage. Silver may be the oldest known anti-bacterial agent. The Ancient Greeks and Romans stored liquids in silver vessels to prevent bacterial growth and spoilage,[11] and the Romans put silver spoons in water to purify it. Silver sutures for wounds have been used from the 1800s. Silver nanoparticles, which are undoubtedly covered in silver oxide are used as antibacterial coatings on wound dressings, biomedical devices, clothing and even keyboards. Samsung, and maybe other appliance makers, coat the interior surfaces of refrigerators and air conditioners with silver nanoparticles to provide both anti-bacterial and anti-fungal effects.[12] It is used in tooth fillings where the silver is dissolved in mercury to form an amalgam. Maybe in that case, the silver also adds an anti-bacterial role.

Silver was used for tableware so much so that cutlery is sometimes referred to as silverware even though nowadays it is much more likely to be stainless steel, which does not tarnish like silver. Plates and cutlery may either be alloyed silver or plated with silver. Silver is also used for jewelry, in which case typically copper is added to harden it since pure silver is too soft, see Fig. 40.2. For instance, Sterling silver contains 7% copper and some rings and bracelets may contain up to 20% copper. Silver slowly gains a black tarnishes over time due to hydrogen sulfide (H_2S) in the air that produces silver sulfide on the surface of the silver.

---

[11] https://frugallysustainable.com/silver-for-water-purification/

[12] http://www.silverdoctors.com/silver/silver-news/samsung-introduces-washing-machines-using-patented-nano-silver-ions/

The silver salts silver chloride (AgCl) and silver iodide (AgI) have always been the basis of film-based photography since they undergo photochemical decomposition and darken. While film-based photography is dying out, these silver salts are still used in photochromic lenses, where copper ions in the glass reverse the darkening.

New uses have continually been found for silver. More recently, it is used in high-quality electronics and various photovoltaic panels and thermal solar reflectors, and high performance glazing as 10–15 nm (one nanometer is a billionth of a meter) coatings.

World silver production was 25,000 tonnes in 2017, with Mexico and Peru producing the most at 5600 tonnes and 4500 tonnes, respectively.[13]

---

[13] USGS, Silver, 2017.

Chapter 41
Steel

There are numerous types of steels and so it is useful to start with a definition of steel. In fact, for simplicity rather than defining steel, let's start by defining *plain carbon steel*. (We should note that modern plain carbon steels, perhaps confusingly, aren't just carbon. They also contain small amounts of other alloying elements, for example up to 1.65% wt. % manganese, and up to 0.6 wt. % of silicon and/or copper). A plain carbon steel is an alloy of iron and carbon with 0.05–2.1 weight percent (wt. %) carbon. Historically, iron-carbon alloys with very low levels of carbon (less than about 0.08 wt. %), and often containing some slag from the melting process, have been called *wrought iron*. Conversely, iron carbon alloys with large carbon contents of 1.8–4 wt.% carbon are called *cast irons*. The monikers *wrought* and *cast* result from the method of processing. Iron has a body centered cubic (b.c.c.) crystal structure at room temperature, which is referred to as ferrite, see Fig. 41.1. Carbon atoms are virtually insoluble in b.c.c. iron, because the carbon atom is too big to easily fit into the holes or interstices in the carbon unit cell. Thus, the carbon that can't dissolve in the b.c.c. iron forms a carbide precipitate, Fe_3C, called cementite, which has a complex orthorhombic crystal structure.

To continue with our definitions: an alloy steel is an iron alloy that contains from 1 to 50 wt. % of other alloying elements. The most common alloying elements are manganese, nickel, chromium, molybdenum, titanium, vanadium, tungsten, cobalt, niobium and boron. In sufficient quantities, these alloying elements can change the equilibrium crystal structure. For example, nickel stabilizes the face-centered cubic (f.c.c.) crystal structure, called Austenite, see Fig. 41.1. The alloying elements chromium and nickel are particularly important and their addition in substantial quantities gives rise to essentially a new class of corrosion-resistant alloys called stainless steels – greater than 11 wt. % chromium is required to make the steels "stainless", that is resistant to corrosion and oxidation. Stainless steels are so important that they are discussed in a separate chapter.

Steels are not new, but their mass production dates back barely 150 years. We may never know when and where steel was first produced, however, steel dating to 1800 B.C. was found at a proto-Hittite site Kaman-Kalehöyük, Kırşehir Province, in

© Springer International Publishing AG, part of Springer Nature 2018
I. Baker, *Fifty Materials That Make the World*,
https://doi.org/10.1007/978-3-319-78766-4_41

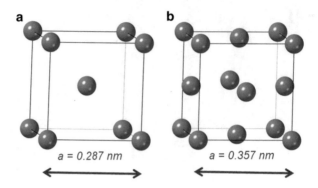

Fig. 41.1 The unit cells of (**a**) b.c.c. iron, and (**b**) f.c.c. iron

modern-day Turkey [1]. Initial diffusion of iron and steel technology throughout the Middle East can be attributed to the Hittites [1]. A driving force for producing steel was the military need for superior swords and for armor. This led to the development of the famous Damascus swords made from high-carbon (1.3–1.8 wt. %) Wootz steel [2–5] that was produced in India from at least the third century B.C. [6]. and to the Japanese long sword, the Katana, which had a hard high-carbon core and softer low-carbon outer edge [7]. During both the Iron Age and later, the best weapons were made from steel since they had much higher strength than wrought iron. However, steel was very difficult to make because there was no way to measure or control the carbon content. Ancient blacksmithing was an art rather than a science and sharp-edged steel swords were developed through trial and error. Wrought iron was always easier to make, and, hence, much cheaper and, thus, much more common.

While technological developments in the production of iron, such as: the use of coke rather than charcoal by Abraham Darby (1678–1717); the development of the Puddling Furnace by Henry Cort (1741–1800); and the development of the hot blast technique by James Beaumont Neilson (1792 –1865) improved the quality and lowered the cost of iron, it was the invention of the Bessemer conver-tor in 1856 that led to the reproducible mass production of steel [8]. Sir Henry Bessemer [9] (1813–1898) developed a pear-shaped furnace in which oxygen blown through the bottom rapidly oxidizes excessive carbon from molten iron: the oxidation of the carbon is quite exothermic and, thus, also heats the furnace. While Bessemer's blast furnace was a great innovation, it still had two prob-lems. First, the process removed too much carbon, rendering the iron quite weak. Second, the process did not remove the phosphorous that was commonly present, and which makes steel brittle. These two problems were solved by ingenious inno-vations by others. First, Briton Robert Forester Mushet (1811–1891) found that by burning off all the carbon and most other impurities, and then adding the right amount of *Spiegeleisen*, an iron alloy containing approximately 15% manganese with small amounts of carbon and silicon, the iron was turned into steel. Later, in 1878, another Briton, Sidney Gilchrist Thomas (1850–1885), added limestone to the molten iron in the Bessemer convertor. This reacted with the phosphorus, which ends up in the slag and can be removed. These innovations produced a dra-matic drop in the price of steel from $170 per ton in 1867 to $32 per ton in 1884.

The result was the use of steel in numerous applications. One obvious one was the complete replacement of iron rails with much more durable steel ones.

While the invention of the Bessemer convertor and its ancillary developments was a disruptive technology, by 1900 it had largely been displaced by the larger capacity Open-Heath Furnace developed by the German-British inventor and engineer Sir Charles William Siemens (1823–1883) [10]. In the 1960s the latter device was itself replaced by the Basic Oxygen Process, an improved version of the Bessemer convertor, developed by the Swiss engineer Robert Durrer (1890–1978) [11] and, for recycled scrap, the Electric Arc Furnace. The introduction of the latter led to the rise of mini-mills – as opposed to large integrated steels mills where ore comes into the plant and finished steel leaves – in the U.S.A. starting with Nucor corp. in 1969. Mini-mill processing ushered in another large reduction in the cost of steel. Remarkably, while the Electric Arc Furnace is the "latest" technology for steel making, it actually isn't new at all. William Siemens in addition to inventing the Open-Heath Furnace, a high temperature pyrometer for measuring furnace temperatures, and speculating on using the sun to power furnaces, also demonstrated a working electric arc furnace in 1879–1880 [9].

The iron-carbon phase diagram, a map of which phase is present at a given temperature and composition, is quite complicated. The one used to describe steels is in fact not the equilibrium phase diagram, which would show graphite instead of Fe_3C. However, for practical purposes Fe_3C is the phase that forms in steels. The development of a particular microstructure in steel starts with heating the steel into the temperature range where iron adopts the f.c.c. (austenite) crystal structure. This temperature range depends on the carbon content – it is 912–1394 °C for carbon-free iron, and as low as 727 °C for iron containing 0.76 wt. % carbon. Carbon is soluble in austenite up to 2.14 wt. % C at 1147 °C because of the large interstitial holes in the crystal structure in which the carbon can sit. The microstructure developed in a steel depends not only on its composition but also on its thermal history, that is what temperatures the material has seen and how quickly the temperature changed. Several different arrangements of ferrite and cementite can be formed, depending on the heat treatment, all of which are given names such as Bainite and Pearlite. Figure 41.2 shows the pearlite structure that is common to many steels, which consists of alternating lamellae of ferrite and cementite. If the high-temperature carbon-containing f.c.c. austenite is rapidly quenched to below room temperature the carbon does not have time to form cementite but is retained in solution and produces a fine lath structure called martensite, which has a body-centered tetragonal (b.c.t.) crystal structure, see Fig. 41.3. Martensite is very similar to ferrite, but the carbon retained in the unit cell shown in Fig. 41.1 causes one of the axes to expand. If this Martensite is heated up (to below the temperature at which Austenite forms) the equilibrium phases form and the resulting microstructure, which consists of cementite particles in a ferrite matrix, is call Tempered Martensite.

The addition of alloying elements affects the microstructures produced even introducing different phases. Complicated heat treatments can be used to produce optimum microstructures. Alloying can be used to affect the strength, toughness, ductility, wear resistance, corrosion resistance, and both the high and low temperature behavior. Although steel is a very old material, it is also a very advanced material. Research continues apace to produce new steels that can be used at higher

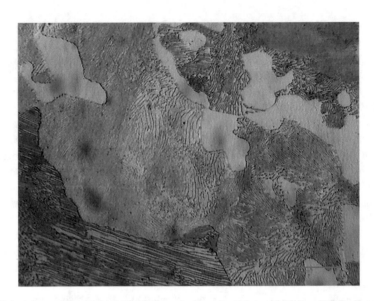

Fig. 41.2 A hypoeutectoid steel. The light regions are b.c.c. iron, which is called ferrite, and the dark regions are Fe₃C, which is called cementite. The regions containing alternating lamellae of ferrite and cementite are called pearlite. The fine structure of the pearlite, which is common to all mild steels, gives the steel its strength. (Courtesy of Daniel Cullen)

Fig. 41.3 Optical Micrograph of martensitic (body-centered tetragonal) iron. The martensite unit cell shown is similar that of to b.c.c. iron but the additional of carbon atoms (black) causes the vertical axis to slightly expand (the two horizontal axes contract a little). Since one axis is now different in length to the others, the unit cell is no longer b.c.c., but is called body-centered tetragonal. (Courtesy of H.N. Han)

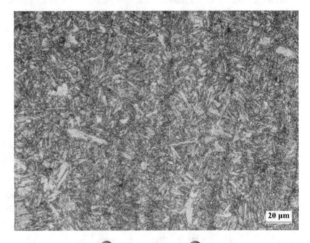

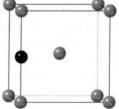

Fig. 41.4 In steel-framed buildings such as this one being constructed at Dartmouth College, the walls are "hung" on the steel frame

temperatures and have higher strength-to-weight ratios. The latter is particularly of interest in the automobile industry where the use of aluminum, which is more corrosion-resistant and much lower density (2700 kg.m^{-3} for aluminum compared to around 7850 kg.m^{-3} for steels) but is more expensive, is making inroads into the use of steel for the bodies. Development of advanced high strength steels (AHSS) has been a hot area of research for automobile applications for many years[1] – you may have noticed the steel on car bodies has become thinner and thinner over the years as higher and higher strength steels, but with essentially the same elastic modulus (200 GPa), are deployed. This means car bodies can bend if you lean on them too heavily. Another major use of steel is in buildings, see Fig. 41.4. The use of steel made the advent of the modern skyscraper possible.

---

[1] Designing Third Generation Advanced High-Strength Steel for Demanding Automotive Applications, Advanced materials and processes, July/Aug 2016.

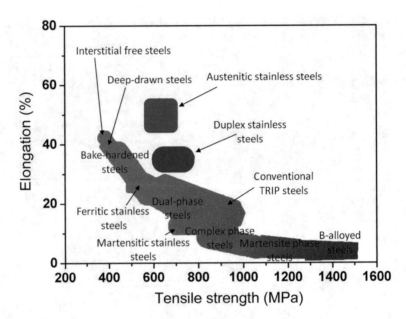

Fig. 41.5 Elongation to failure versus stress at fracture (tensile strength) for a wide variety of steels. (Courtesy of Zhangwei Wang)

Before leaving the mechanical behavior, we should note that as with many things, there are trade-offs. This is demonstrated for steels in Fig. 41.5, where it can be seen that as the tensile (fracture) strength of steels increases their elongation to fracture generally decreases – a classic trade-off in many materials.

The general trend since steel has been mass-produced is for the amount of steel produced to always increase with time. Since 1950 there have been two periods of rapid growth from 1950 to around 1980, and from around 2000 to 2014 to 1612 million tonnes with a value of nearly $1.3 trillion,[2] see Fig. 41.6. The first rise is associated with the post-World War II recovery, while the latter rise is associated with the rise in production in China, which in 2014 accounted for 46% of Global production. Unfortunately, there has been Global overcapacity in the last few years due to overproduction in China, due to its maturing economy. Thus, in 2015 production was 1695 million tonnes whilst consumption was only 1546 million tonnes.[3] Even in April 2017, China was still increasing production.[4] Nevertheless, with the rise of Brazil and India, and other heavily-populated underdeveloped countries, growth of steel demand is inevitable.

Although steel is an advanced material whose chemistry is carefully controlled during production, mild steel only costs around $600/ tonne. This is 60¢/kg or 27¢/ lb. At least in the U.S.A. there is nothing you can buy in the supermarket for as little as 27¢/lb.

---

[2] World+Steel+in+Figures+2016.pdf

[3] http://www.prnewswire.com/news-releases/steel-market-forecast-2015-2025--future-opportunities-for-leading-companies-300108061.html

[4] Materials World, June 2017. P14.

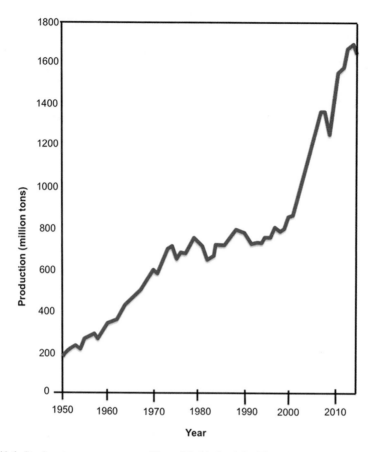

Fig. 41.6 Steel output versus year. (From World+Steel+in+Figures+2016.pdf, https://www.worldsteel.org/)

References

1. Akanuma, H. (2005). The significance of the composition of excavated iron fragments taken from stratum III at the site of Kaman-Kaleh.yük, Turkey. *Anatolian Archaeological Studies, 14*, 147–158.
2. Wadsworth, J. (2015). Archeometallurgy related to swords. *Materials Characterization, 99*, 1–7.
3. Srinivasan, S. (1994). Wootz crucible steel: A newly discovered production site in South India. *The Institute of Archaeology, 5*, 49–59.
4. Biswas, A. R. (1994). Iron and steel in pre-modern India – A critical review. *Indian Journal of the History of Science, 29*, 579–610.
5. Bronson, B. (1986). The making and selling of Wootz, a crucible steel of India. *Archeomaterials, 1*, 13–51.
6. Srinivasan, S., & Ranganathan, S. (2004). *Wootz steel: An advanced material of the ancient world*, Department of Metallurgy, Indian Institute of Science Bangalore. http://materials.iisc.ernet.in/~wootz/heritage/WOOTZ.htm.
7. Chaline, E. (2012). *Steel, fifty minerals that changed the course of history*. Buffalo: Firefly Books ISBN: 13: 978-1-55407-984-1.

8. Spoerl, J. S. *A brief history of iron and steel production*. Saint Anselm College. http://www. anselm.edu/homepage/dbanach/h-carnegie-steel.
9. Amato, I. (1998). *Stuff: The materials the world is made of*. New York: Avon Books, Inc ISBN-10: 0380731533.
10. Carr, J. C., & Taplin, W. (1962). *History of the British steel industry*. Cambridge: Harvard University Press.
11. Smil, V. (2006). *Transforming the twentieth century: technical innovations and their consequences* (Vol. 2). Oxford: Oxford University Press US ISBN 0-19-516875-5.

Chapter 42
Stainless Steel

To quote from the 1949 book "Stainless steels: an elementary text for consumers [1]" by Carl Andrew Zapffe (1912–1994) "'Stainless steel' is not an alloy - it is the name inherited by a great group of alloys, a special classification of special steels, and a field of study in itself." The key difference between stainless steels and other steels is that the former has a few nanometer layer of chromium oxide on the surface while the later form iron oxides. The chromium oxide is adherent and protective against further oxidation and corrosion by many chemicals, but not against chloride attack, whereas iron oxide, which we call rust, is not protective. At least 11% chromium is needed to confer corrosion resistance and produce a stainless steel. The corrosion resistance is enhanced by the addition of some other alloying elements, such as nickel and molybdenum.

One can subdivide stainless steels into five groups of chromium-containing steels or iron-carbon alloys based on their crystal structures see Fig. 42.1, and whether particles are present. These are body-centered cubic (b.c.c.) called ferritic, face-centered cubic (f.c.c.) called austenitic, duplex (a mixture of regions b.c.c. and f.c.c.), body-centered tetragonal called martensitic, and precipitation-hardening (f.c.c. or b.c.t. containing particles). These have different chromium contents and different additional alloying elemenst and, thus, have different mechanical properties and different corrosion resistance [2–4].

Ferritic stainless steels are single-phase alloys that have the same b.c.c. crystal structure as plain carbon steels and thus, as with plain carbon steels they are ferromagnetic. Depending on the exact alloy, they contain from 11–27 wt. % chromium with small amounts of carbon (0.01–0.20 wt. %) and small amounts (less than 2%) of any other elements such as manganese, nickel, silicon or titanium. These stainless steels have relatively low yield strengths for steels of 205–350 MPa and elongations of 20–25%. Their yield strength can be improved by about 50% by cold work with only a little loss in ductility. Typical uses include: kitchen equipment, automobile trim and exhausts, high temperature valves and combustion chambers.

© Springer International Publishing AG, part of Springer Nature 2018
I. Baker, *Fifty Materials That Make the World*,
https://doi.org/10.1007/978-3-319-78766-4_42

Fig. 42.1 The crystal structures of (**a**) b.c.c. iron (ferrite), (**b**) f.c.c. iron (austenite), and (**c**) b.c.t. iron (martensite). In the tetragonal crystal structure of martensitic steel the vertical axis is longer than the two horizontal axes, which are around $a = 0.287$ nm, and the c/a ratio depends on the carbon content of the alloy

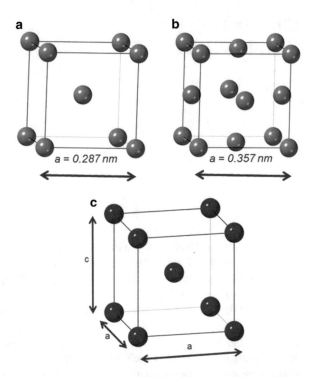

Austenitic stainless steels are single-phase alloys that have the f.c.c. crystal structure that is adopted by unalloyed iron above 912 °C. This crystal structure is stabilized by the presence of typically 7–12 wt.% nickel and manganese and may contain small amounts other elements such as titanium, silicon or molybdenum. These stainless steels contain small amounts of carbon (less than 0.15 wt. %) and 17–19 wt. % chromium. Austenitic stainless steels, which are not magnetic, account for 70% of stainless steel production and include the well-known 18/8 (18% chromium, 8% nickel). Austenitic steels are more corrosion resistant than ferritic steels because of the high nickel and chromium contents, and, thus, are usually more expensive. They are generally not as strong as ferritic steels exhibiting yield strengths of 170–290 MPa, but they show greater ductility with elongations of 40–60%. Like ferritic stainless steels, they can also be strengthened considerably by cold working such as rolling, which can more than double the yield strengths but with a substantial reduction in ductility. Because austenitic stainless steels have a f.c.c. crystal structure they do not become brittle at low temperature and can be used for cryogenic applications – conversely ferritic stainless steels like all b.c.c. iron alloys become brittle at low temperature. Austenitic stainless steels are used for many structural applications, in transportation, for components that need welding and various chemical tanks.

Duplex stainless steels attempt to combine the desirable properties of ferritic (b.c.c.) and austenitic (b.c.c.) stainless steels and have a microstructure that is roughly a 50:50 mix of alternating crystals or grains of these two phases.[1] They typically have high chromium contents of 19–32 wt. %, which confers excellent corrosion resistance, lower nickel contents than austenitic stainless steels and up to around 3.5 wt. % molybdenum or manganese. They are typically about twice as strong as austenitic steels at 400–550 MPa, although not as tough, and the lower nickel content (less than 7 wt. %) means that they are more affordable. They are used in applications such as swimming pool fixtures, brewing tanks, hot water tanks and pressure tanks.

Martensitic (b.c.t.) stainless steels contain 12–18 wt. % chromium and less than 1 wt. % manganese and molybdenum and up to 1 wt. % carbon, but little nickel.[2] Thus, they are cheaper than the austenitic stainless steels, but do not have the corrosion resistance. However, they are considerably stronger. Historically, these were the first commercial stainless steels having been invented by the English metallurgist, Harry Brearley (1871–1948) [5] in 1913 when developing alloys for corrosion-resistant gun barrels. Martensitic stainless steels have to be heat treated in a similar way to plain-carbon steels. The steel is heated to a temperature where the high-temperature austenite (f.c.c.) crystal structure is stable and then cooled rapidly to produce the body-centered tetragonal martensite phase, see Fig. 42.1, and then tempered to reduce the stresses in the material. This produces a very high strength, which, depending on the exact alloy, can be up to 2000 MPa, but in this condition the stainless steel shows elongations as little as 5% – untempered martensitic stainless steels have low toughness and are brittle. Martensitic stainless steels are used for medical applications (scalpels, clamps), cutlery, bearings, valves, springs and pump shafts.

Precipitation-hardenable stainless steels have typically 16–17 wt. % chromium and 4–7 wt. % nickel with up to a few percent of other elements such as aluminum, copper, molybdenum and manganese. They can be martensitic (b.c.t.), and austenitic (f.c.c.) or austenitic-martensitic. These steels have comparable corrosion resistance to austenitic stainless steels but can be heat treated to produce precipitates that considerably strengthen the steel. Room temperature yield strengths of around 1700 MPa are possible but with typical elongations of only 1–6%.[3] These stainless steels are used in cams, gears, airframe parts, turbine parts and knives.

The development of stainless steel was not an overnight success story but followed a century of efforts by various individuals. Following the isolation of chromium in 1780 by the French chemist Louis Nicholas Vauquelin (1753–1829) [6], several researchers investigated iron-chromium alloys including the famous English metallurgist Sir Robert A. Hadfield (1858–1940). The latter concluded that chromium worsens the corrosion resistance of steel. This seemingly odd conclusion

---

[1] http://www.imoa.info/molybdenum-uses/molybdenum-grade-stainless-steels/duplex-stainless-steel.php

[2] http://www.aksteel.com/markets_products/stainless.aspx

[3] http://www.twi-global.com/technical-knowledge/job-knowledge/precipitation-hardening-stainless-steels-102/

arose because he performed tests in sulfuric acid, to which stainless steels are not resistant. However, no one came up with a useful stainless steel because either they produced steels in which the chromium content was too low or the carbon content too high. The latter leads to carbide formation, which locally depletes the chromium level and worsens the corrosion resistance [7] – a similar problem can occur in some modern stainless steels when they are welded. The invention of stainless steel had to wait until the twentieth century and the work of the French professor of metallurgy and metal processing Léon Guillet (1873–1946). Guillet published a paper on the microstructures and mechanical properties of alloys that are now considered typical ferritic and martensitic alloys starting in 1904, and on essentially austenitic stainless steels starting in 1906. Surprisingly, Guillet appears not to have filed any patents or studied the corrosion resistance of these new steels. It fell to the German Philipp Monnartzin in 1911 to publish the first work showing that adding 12% chromium to iron greatly increased the corrosion resistance. Research on stainless steels was now starting to heat up in the U.K., France, Germany and the U.S.A. A huge step forward was when Harry Brearley (1871–1948) working at the Brown-Firth Research Laboratory in Sheffield, England produced the first commercial stainless steel, an alloy containing 13 wt. % chromium and 0.24 wt.% carbon that is very close to the composition of modern martensitic stainless steels. His development led to Sheffield, a city that had been the center of cutlery production outside London since 1600 A.D., becoming the residence for the new stainless steel cutlery industry.[4] Of course, Brearley was not alone in the development of stainless steel. Around the same time: the American Elwood Haynes (1857–1925) was developing martensitic stainless steels – Haynes is more well-known for his invention of the wear resistant cobalt-chromium Stellite alloys and for starting the Haynes Stellite company; the Americans Frederick M. Becket (1875–1942) and Christian Dantsizen were developing ferritic stainless steels; and the German Eduard Maurer (1886–1969) working in the Friedrich A. Krupps research laboratories in Essen, Germany demonstrated the corrosion resistance of austenitic stainless steels.

The utility of stainless steels for other uses than cutlery was soon recognized. The nascent Royal Air Force used stainless steel in the exhaust valves of aircraft engines and in 1916 the British government prohibited the use of stainless steels for anything other than defense purposes until the end of the First World War. Stainless steels enabled another important defense-related innovation: the first materials used for the disks and blades in Sir Frank Whittle's (1907–1996) jet engine were various austenitic stainless steels including Stayblade steel, Firth-Vickers Rex 78 and G18B steel. However, because of their low strength at high temperatures stainless steels limited the operating temperature and it was not long before superior nickel-based superalloys that could operate at higher temperatures were developed [5].

While the main classes of stainless steels were invented over a century ago, a significant breakthrough in their processing was the use of the argon-oxygen decarburization (AOD) process. This was invented in 1954 by the U.S. company Union

---

[4] http://www.localhistories.org/sheffield.html

Fig. 42.2 Stainless steel is mostly commonly encountered in saucepans and cutlery

Carbide Corp., renamed Praxair after 1992.[5,6] Scrap or virgin materials are initially melted in an electric arc furnace, after which the stainless steel is subjected to the AOD process in which an argon/oxygen or nitrogen/oxygen mixture is injected into the stainless steel. This process removes carbon to levels less than 0.05% by oxidizing it to carbon monoxide, but avoids oxidizing and, hence, losing expensive chromium. It has the additional advantage that inexpensive raw materials, such as high-carbon ferrochromium, can be used as feedstock. The process is also used to make some tool steels, silicon steels and cobalt-base and nickel-base alloys.

Stainless steels are well-developed materials that have found many uses from toasters to trains to cars. The DeLorean DMC-12, a spectacular, if commercially unsuccessful, gull-winged car produced in Northern Ireland by the DeLorean Motor Company from 1981–1983, had brushed stainless steel body panels. It is actually not clear why one would want to make a car body that will last a very long time – the engine power and efficiency, braking systems, safety systems, and electronics in a DeLorean look very dated compared to a modern car. Stainless steel is *de rigueur* in modern restaurant kitchens and in surgical instrument use, see Fig. 42.2. A list of all the uses of stainless steels would be very lengthy: it finds uses where corrosion and oxidation resistance is needed.

Some iconic structures are made of stainless steel: the 192 m high Gateway Arch in St. Louis, Mo. is clad in 804 tonnes of 304 stainless steel; the arc deco terraced crown of the 282 m high Chrysler building in New York is clad in stainless steel, and of more recent construction the 379 m twin Petronas Towers in Kuala Lumpur, Malaysia are covered in 83,500 square meters of stainless steel. Stainless steel production is likely to continue and expand its use in such prestige structures as well as for many other applications.

---

[5] http://www.praxair.com/industries/metal-production/argon-oxygen-decarburization-aod

[6] https://www.britannica.com/technology/argon-oxygen-decarburization

Apart from in 2008 and 2009 Global stainless steel production has increased each year, and from 2005 to 2016 it grew from 25 million tonnes to 46 million tonnes. This is a fraction of total steel production, which was 1630 million tonnes in 2016. One reason that stainless steel is not used more is that is costs $2200/ tonne (304 coil) whereas mild steel is around $600/ tonne in 2017.

References

1. Zapffe, C. A. (1949). *Stainless steels: an elementary textbook for consumers.* American Society for Metals, Metals Park, Cleveland, OH. pp. 349.
2. Flinn, R. A., Paul, K., & Trojan. (1995). *Engineering materials and their applications.* Hoboken, NJ: Wiley ISBN-13: 978-0471125082.
3. Smith, W. F. (1990). *Principles of materials science and engineering.* New York: Mcgraw Hill Publishing Company ISBN-13: 978-0070592414.
4. William, D., & Callister. (2001). *Fundamentals of materials science and engineering: An interactive e . Text.* New York: Wiley ISBN-13: 978-0471395515.
5. Cobb, H. M. (2010). *The history of stainless steel.* Materials Park: ASM International ISBN 13: 978-1-61503-011-8.
6. Emsley, J. (2001). *Nature's building blocks: An A–Z guide to the elements.* Oxford: Oxford University Press ISBN: 0-19-850340-7.
7. Koff, B. L. (2004). Gas turbine technology evolution: A designers perspective. *Journal of Propulsion and Power, 20*(4), 577–595.

Chapter 43
Stone

The Stone Age was the long epoch in human history that ended around the fifth millennium B.C. with the smelting of metals in the Fertile Crescent, and elsewhere somewhat later. We can date the start of the Stone Age to around 3.3 million years ago, the date of the oldest known stone tools found in Kenya. Oddly, these particular tools appear to predate any evidence of humans [1, 2]. This suggests either that humans weren't the first to use stone tools or that humans existed before the oldest evidence of their existence that has been found to date. The oldest stone tools that were clearly used by humans date to 2.6 million years ago, and were found in Ethiopia. Stone tools were made from a variety of materials including a number of silica-based rocks and minerals such as the sedimentary rocks chert or flint and radiolarite; the mineral chalcedony; basalt, a volcanic rock which is 65% feldspar (complex silicates); and metamorphic quartzite, see Fig. 43.1. Objects were also made from Obsidian an igneous natural glass produced in volcanoes. Obsidian is quite brittle and easily fractures producing sharp edges that can be used for cutting [3]. In fact, Obsidian is still used today to make surgical knives because of the very sharp edge that can be produced [4].

Many stone tools, including axe-heads, chisels and polishing tools, were made by striking stone flakes from a stone, see Fig. 43.2. Hand-axes could be made by joining these stones to wooden handles using natural glue such as bitumen or tree resin. The hand-axes and other stone tools evolved and became more sophisticated over time. Some jewelry was also made of stones, particularly Obsidian. The use of stone not only provided humans with tools but their use has been linked to evolution. Kenneth Page Oakley, a British anthropologist, (1911–981) in his book "Man the Tool-maker" promoted the idea that stone tool-making propelled the evolution of our "powers of mental and bodily co-ordination" [5]. It has also been suggested that the processes of both teaching and learning complex tool making skills may have contributed to the development of language in humans [6].

Dry stone wall construction, a technique that does not use any binding material, dates back to the Neolithic age. It was used to construct buildings and for fencing enclosures. Stone buildings are relatively easy to construct. They have the advan-

© Springer International Publishing AG, part of Springer Nature 2018
I. Baker, *Fifty Materials That Make the World*,
https://doi.org/10.1007/978-3-319-78766-4_43

Fig. 43.1 A collection of stone cutting tools in the Otago Museum, Dunedin, New Zealand

Fig. 43.2 A greenstone
axe head in in the Otago
Museum, Dunedin, New
Zealand

tage that they are not only durable, but also are rot proof, and bug and vermin proof.
Cupboards, dressers and beds made of stone can be found in the stone-walled buildings at a near-intact Late Neolithic village called Skara Brae located on the Mainland,
Orkney, Scotland, which was inhabited around 3200–2200 B.C. for about 600 years
[7][1] see Fig. 43.3. Some famous structures from antiquity survive because they are
made of stone such as the numerous pyramids in Egypt and the Neolithic stone
circles at Stonehenge and Avebury in England. While stone tools have long been
superseded, the use of dry stone walls to enclose fields continues to this day as does
the use of stone for building. Limestone wall-enclosed fields were very common in
the north of the county of Derbyshire, England where I grew up, and my garden has
the remains of a granite stone wall, see Fig. 43.4.

---

[1] http://www.orkneyjar.com/history/skarabrae/

Fig. 43.3 Furniture made of stone inside a Neolithic stone-walled dwelling at Skara Brae located on the Mainland, Orkney, Scotland

Fig. 43.4 The remains of an old stone wall that borders the author's garden

Fig. 43.5 Part of the Great Wall of China neat Bejing

Stone has been a leading building material throughout history. It was used to build palaces, castles, religious buildings and important government buildings from ancient times to the present. The Ancient Egyptians, who constructed large pyramids and temples starting around 5000 years ago, initiated the trend for building large structures out of stone. The Great Pyramid of Giza (or Cheops), constructed largely of huge stones blocks was completed in the twenty-sixth century B.C. and was the tallest building in the world at 481 m for 3800 years. The Ancient Greeks continued this tradition of building temples out of stone. The Romans used marble and granite extensively for buildings, but also constructed many buildings, such as the Coliseum, out of concrete, some of which were faced with stone. The Romans made extensive use of stone for road building such as the 444 km long Watling Street in England that was built on ancient pathways. The Romans also built defensive structures out of stone such as the 118 km long Hadrian's Wall constructed in 122 A.D. when Hadrian was the Roman Emperor. It lies in England not too far from the border with Scotland – 160 km further north is the 63 km long Antonine Wall constructed in A.D. 142 that was simply a wood and turf fortification. The Chinese built one of the most famous structures in the world mostly out of stone, the Great Wall of China, See Fig. 43.5. The Great Wall is in fact not one wall but many walls that were built to prevent invasions from the north starting in the second century B.C. and not ending until the seventeenth century A.D. All together, there are over 21,000 km of wall. During the Middle Ages, castles and churches were built all over Europe, with some masterpieces in the form of huge cathedrals. The construction of these structures included features such as the Gothic arch and flying buttresses, whose development

Fig. 43.6 Granite flooring. Ironically, this particular granite, which is very commonly used around the world, is from China and is installed in the Thayer School of Engineering, Dartmouth College New Hampshire, whose nickname is the Granite State

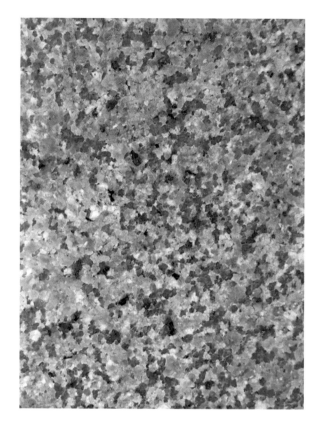

had started centuries earlier, but whose use blossomed during the 12–16th centuries. Meanwhile in Asia, magnificent constructions were also being built from stone including the twelfth century Angkor Wat temple complex in Cambodia.

Modern uses of stone include constructing buildings and walls, decorative facing on buildings and flooring. It is even sometimes used as outdoor furniture such as stone benches. Stone's resistance to weathering and wear, high strength and its aesthetic appeal are important considerations in many applications. For example, granite, an igneous rock, is commonly used for flooring (see Fig. 43.6), countertops, and some buildings; sandstone, a sedimentary rock, is used for buildings such as the quadrangle buildings at the University of Sydney, Australia; marble, a metamorphic rock, is used for its strength and beauty in buildings, the most famous use of which is the seventeenth century marble-clad Taj Mahal in Agra, India; and slate, a fine-grained, sheet-like metamorphic rock, is often used for roofing tiles.

In addition to its use as large chunks, stone is also used as stone dust and crushed rock is used for driveway construction and for a base layer underneath either paving slabs, which may be made of stone, or asphalt, which is stone coated with bitumen. Stone powder can also be used instead of sand in concrete.

Thus, while stone is an ancient and low-tech material some of its outstanding properties mean that its use will continue far into the future.

References

1. Morelle, R. *Oldest stone tools pre-date earliest humans.* http://www.bbc.com/news/science-environment-32804177
2. Wong, K. The new origins of technology. *Scientific American*, May 2017. pp. 28–35.
3. Amato, I. (1998). *Stuff: The materials the world is made of.* New York: Avon Books, Inc ISBN-10: 0380731533.
4. Buck, B. A. (1982). Ancient Technology in Contemporary Surgery. *The Western Journal of Medicine, 136*, 265–269 PMID 7046256.
5. Oakley, K. (1972). *Man the tool maker.* London: Natural History Museum Publications ISBN: 13: 9780565005382.
6. Stout, D. Tales of a stone age neuroscientist. *Scientific American*, April 2016. pp. 29–35.
7. Magazine 1843, April and May issue 2017, p. 108.

Chapter 44
Tin

Historically, the most important use of tin was to alloy it with copper to produce bronze. Tin is easily melted since it has a melting point of only 232 °C. Bronze was used as early as 3500 B.C. in present-day Turkey and ushered in a new age [1]. However, neither copper nor tin are very abundant in the Earth's crust at 0.0068% and 0.00022%, respectively,[1] and they are rarely found together. In the Ancient Western World, principally south-west England, in what would become the county of Cornwall, was a prime source of tin as early as 2150 B.C. and mining continued there until 1998 [2] – tin was also found in northern Spain and Brittany, France. As tin sources in the Middle East were exhausted, Cornish tin became an important source. Tin is easily extracted from its principal ore, Cassiterite, which is tin oxide, by heating with carbon [2]. In 2014, over 70% of the current World's tin production (296,000 tonnes) comes from just two countries China (125,000 tonnes) and Indonesia (84,000 tonnes).[3]

Although tin is corrosion resistant and malleable, it is too soft and expensive at ten times the price of aluminum [4] to be used as a structural material. From at least 1450 B.C. the Ancient Egyptians used tin to produce the alloy pewter.[5] Traditionally, pewter contained 85–90% tin with varying amounts of copper, lead and antimony depending on the exact alloy: modern pewter contains bismuth rather than lead. Pewter, a dull gray metal that can look bluish at higher lead concentrations, has an even lower melting point than tin at 170–230 °C and is easily malleable [3].[6] From the twelfth to nineteenth centuries it was used extensively in Europe for tableware, cutlery, drinking cups and food containers some of which had elaborate designs.

---

[1] http://periodictable.com/Properties/A/CrustAbundance.an.html

[2] https://www.croftytin.co.uk/history-of-cornish-tin-mining/

[3] http://www.worldatlas.com/articles/leading-tin-producing-countries-in-the-world.html

[4] https://www.fastmarkets.com/base-metals/tin-prices-and-charts/

[5] https://www.britannica.com/technology/pewter

[6] http://www.ramshornstudio.com/pewter.htm

© Springer International Publishing AG, part of Springer Nature 2018
I. Baker, *Fifty Materials That Make the World*,
https://doi.org/10.1007/978-3-319-78766-4_44

Fig. 44.1 A "tin can",
which is corrugated for
strength. Many "tin cans"
don't have a tin coating,
but use a cheaper plastic
coating over the steel

With the advent of cheap glassware and porcelain, the use of pewter for these purposes eventually disappeared. Pewter has limited use currently and is mainly used for various decorative or collectible objects.

In the nineteenth century, tin found a major use in tin cans. Peter Durand, a British merchant, obtained a patent for preserving food in tin cans in 1810, which he sold to two other Britons, Bryan Donkin and John Hall, who set up the first commercial tin canning factory in London in 1813.[7] Tin does not react with food and so tin cans enabled food to be preserved for long periods. This was of great use for the military and for long distance sea journeys where previously food had to be dried, smoked, or cured with salt. Tin cans were first used by the Royal Navy. These early tin cans were sealed using a lead-tin alloy, which was not ideal for the diner's long-term health. Surprisingly, the invention of the can opener did not occur until 1855 when Robert Yeates, an English maker of cutlery and surgical instruments, developed one.[8] One wonders what happened to all the tin cans before the can opener was invented.

Tin cans are actually steel with a tin coating that protects the steel from corrosion – a technology that dates from Ancient times. Tinplate was also used for pots and pans from the early seventeenth century onwards, a use now discontinued. Many "tin cans" now have no tin but use a cheaper plastic coating on steel see Fig. 44.1.

Tin occurs in two different crystal structures, white tin, a silvery gray ductile metal with a body-centered tetragonal crystal structure, see Fig. 44.2, that is the stable above 13.2 °C and gray tin, which is a gray diamond-cubic structured brittle powdery material, see Fig 44.2, that is stable below this temperature. Upon cooling, the transformation from white tin to gray tin involves a huge volume expansion of 27% [4]. This expansion causes stresses and failure of the material. Historically, this has had important consequences. During World War II, the tin in German soldiers canned food disintegrated, thus, contaminating their food during the Winter campaign against Russia. This disintegration is referred to as "tin pest". The failure of organ pipes in Russia during the cold winter of 1850 has also been attributed to this process.

---

[7] https://www.britannica.com/topic/canning-food-processing#ref135266

[8] https://targetstudy.com/knowledge/invention/137/can-opener.html

Fig. 44.2 Crystal structure of (**a**) body-centered tetragonal crystal white tin, and (**b**) diamond-cubic gray tin. For gray tin all the sides of the unit call have the same length. White tin is the stable allotrope above 13.2 °C and gray tin is stable below that temperature

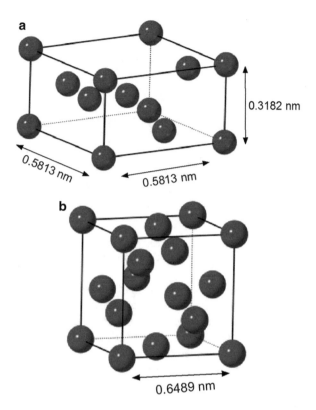

An interesting feature of tin and some other metals such as niobium, indium, zinc, cadmium, gallium, and mercury is twinning cry.[9] When a piece of tin is bent sheets of atoms move over each other in a process referred to as twinning, see Fig. 44.3, which produces a rapid series of clicks that sound like crying.

As evident from above, the use of tin has been declining, but there are still key uses for tin. An important modern use is in the float glass process whereby window glass of any desired thickness is made by floating molten glass on a bed of molten tin under a protective atmosphere. Tin is used for this purpose because of its low melting point, low toxicity and because it reacts little with glass.[10]

Tin has added to the vernacular in various pejorative uses. "Tinny" is used to mean poor-quality in a metallic object; disparagingly regarding music; and negatively regarding the taste of food. Similarly it is used in the expression "tin ear", which refers to not only not being able to appreciate music, but also figuratively for not listening properly or understanding what people are saying.

---

[9] http://www.sharrettsplating.com/blog/the-tin-plating-process-a-step-by-step-guide/

[10] http://www.pilkington.com/pilkington-information/about+pilkington/education/float+process/step+by+step.htm

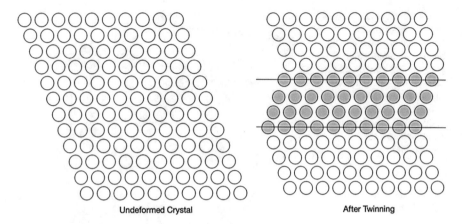

Undeformed Crystal   After Twinning

Fig. 44.3 The twinning process in tin occurs by the co-operative movement of many atoms during deformation and produces a so-called twinning cry in tin

Tin can be said to hold much of the world together. Solders have been typically made of either 60% tin/40% lead or 63% tin/37% lead for joining electronics or 50/50 tin/lead for plumbing. The solders used for electronics melt at 183–187 °C, which is much lower than the melting temperatures of the metals being joined. Lead solders are now banned from use in plumbing. In July 1, 2006, the European Union issued a directive to reduce lead in electronics, which has spurred development of lead-free solders. These newer solders still use tin and are based on tin-silver and copper alloys [5]. Some manufacturers have replaced lead-tin solders with pure tin opening up the possibility of the return of so called "tin pest", but this can be mitigated by alloying with antimony or bismuth, which prevent the decomposition. Another issue with using tin in electronics is the formation of tin whiskers due to compressive stresses that can arise. [11] The whiskers can be a millimeter long or more and can cause short circuits, break off and damage disk drives or even act as miniature antennae.

2016 World tin production was around 296,000 tonnes, with China and Indonesia accounting for 125,000 tonnes and 84,000 tonnes, respectively, or together 70% of World production. The amount of tin produced each year is miniscule compared to steel production at 1630 million tonnes (2016). The low production of tin is because of its scarcity and cost. At $21,000/tonne in February, 2018, or $21/kg it is hugely expensive compared to steel which is around 60¢/kg.

---

[11] BBC podcast TIN, March 1st, 2014.

References

1. Stwertka, A. (2012). *A guide to the elements.* New York: Oxford University Press ISBN: 978-0-19-983252-1.
2. Chaline, E. (2012). *Fifty minerals that changed the course of history.* Buffalo: Firefly Books Ltd.
3. Hull, C., *Pewter.* (1992). Oxford: Osprey Publishing. ISBN 978-0-7478-0152-8.
4. Schaffer, J. P., Saxena, A., Antolovich, S. D., Sanders, T. H., Jr., Warner, S. B., & Warner, S. B. (1999). *The science and design of engineering materials.* Taipei: McGraw-Hill ISBN 0-256-24766-8.
5. Ganesan, S., & Pecht, M. (2006). *Lead-free electronics.* Hoboken: Wiley ISBN 0-471-78617-9.

Chapter 45
Titanium

Titanium can be considered a god among metals. Like the Titans of Greek mythology, deities of incredible strength, alloys made of titanium are extremely durable and can be very strong. Thus, in 1795 the Austrian chemist Martin Klaproth (1743–1817) named the metal after these deities. Like most pure metals, the pure silvery gray pure metal is relatively soft, but titanium alloys such as Ti-3%Al-8%V-6%Cr-4%Zr-4%Mo can be very strong (yield strength of 1100 MPa) and because of the relatively low density of titanium (4500 kg/m^3), they have a much better specific strength (strength/density) than many steels. For example, Ti-10%V-2%Fe-3%Al has a specific strength of 264 kN.m/kg while a typical 18/8 stainless steel is only 68 kN.m/kg.[1] Titanium and its alloys are also resistant to oxidation and corrosion. Perversely, the resistance to degradation arises because titanium is very reactive and rapidly forms adherent titanium dioxide, TiO_2, which prevents further environmental attack on the underlying metal.

Because of their excellent corrosion resistance it is useful to compare titanium alloys to stainless steels, which people encounter in everyday life. Titanium alloys are four to five times stronger, 40% lighter and much more corrosion resistant than stainless steels, particularly to chlorides such as salt (NaCl). If titanium alloys are so outstanding compared to stainless steel, why aren't they used more than stainless steel? The answer is, of course, cost. Titanium is ~$10/kg whereas stainless steel is $2–3/kg. The high price is not because titanium is rare. Titanium is the ninth most abundant element in the Earth's crust (0.63% by weight), the seventh most abundant metal, and the fourth most abundant structural metal. The high cost arises because one of the properties that makes it so useful as a structural metal - its rapid formation of a tenacious oxide - is also its Achilles heel (another Greek legend). It is very difficult to win metallic titanium from the oxide-containing ores from which it is derived. The main titanium ores are rutile, which is ~95% titanium oxide (TiO_2), ilmeite ($FeTiO_3$), and leucoxene (a variant of ilmeite of no exact composition in which some of the iron has been leached away). Such ores are currently obtained from alluvial and volcanic deposits mostly in Australia, South Africa and the U.S.A.

---

[1] http://www.azom.com/article.aspx?ArticleID=1341#_Tensile_Strength

© Springer International Publishing AG, part of Springer Nature 2018
I. Baker, *Fifty Materials That Make the World*,
https://doi.org/10.1007/978-3-319-78766-4_45

Fig. 45.1 Knee implant.
The lower metal part is
made of Ti-6Al-4 V; the
white insert is
polyethylene; and the
upper metal part is made of
CoCrMo. (Courtesy of
D.W. Van Citters)

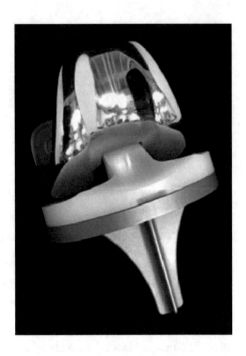

Titanium was discovered in 1791, though identifying an oxide of a previously unrecognized metal, by an English clergyman William Gregor (1761–1817), who was an amateur chemist. It wasn't until 1910 that the American Matthew Hunter (1878–1961) produced metallic titanium by heating titanium chloride ($TiCl_4$) with sodium under great pressure at 700–800 °C. This procedure was later named the "Hunter process". The current process to make titanium, which was patented by Luxembourger William Kroll (1889–1973) in 1940, involves reacting ores containing titanium with chlorine at 800 °C to form a liquid $TiCl_4$, which is then reacted with magnesium to liberate the titanium. The end product is a titanium sponge, which then must be melted under vacuum to prevent oxidation.

Titanium alloys, such as Ti-6 at. % Al- 4 at. % V (often referred to as Ti-6-4), have a number of important applications including in aircraft (half of the titanium alloys used in aircraft are Ti-6-4), in desalination and chemical processing plants, in the oil and gas industry (particularly offshore applications), and in objects like high-performance bicycles (Ti-3Al-2.5 V is often used in this application as well) or golf clubs, in jewelry, and eyeglass frames and watches. It is also used in its pure form as fine particles as a catalyst in some chemical reactions. It has even been used to clad some spectacular buildings such as Bilbao's Guggenheim museum and the Glasgow Science Center. Another important use is in dental applications, and in femoral and knee implants, where its strength, but more particularly its corrosion resistance are important. Its corrosion resistance means that it does not react with the body, which has been an issue in cobalt-chromium alloys or cobalt steels that it has largely replaced in this application, see Fig. 45.1. Titanium is not only biocompatible it is also compatible with examination in a Magnetic Resonance Imaging machine (MRI) and an x-ray computed tomography (X-ray) CT scanner. While implants are a very important application, implants account for only around 1% of the usage of titanium.

Titanium has a long history in the aerospace industry; the first aircraft to use major components of titanium in the airframe was the Douglas X-3 Stiletto in 1952. The Lockheed SR-71 Blackbird spy plane, which first flew in 1964, was the first plane to use mostly titanium alloy construction because of its high strength-to-weight ratio and high strength at high temperatures. This high temperature capability was necessary because of the high temperatures generated on its wings and the front of the aircraft, which reach nearly 500 °C in places, due to the high speed at which it could travel - the plane established the fastest speed for an air-breathing aircraft in 1976 of 3529 km/h. Aluminum could not fulfill this role since it melts at 660 °C compared to 1812 °C for titanium. Starting in the 1960s with planes such as the Boeing 707 and Douglas DC-8 titanium started to be used in civilian aircraft. Titanium's use in the fan disks and blades and the compressor disc and blades in the "cold" section of gas turbine engines started in military aircraft in the 1950s and in civilian airliners in the 1960s. More recently, the Boeing 777 uses 5900 kg of titanium alloy in its landing gear, while the Air Force's F-22 uses 4100 kg of titanium alloy in its airframe; the Airbus A380 uses 63.5 tonnes of titanium, and the Boeing 787 is 15–17% titanium alloy. Two-thirds of metallic titanium is used in aircraft. It is also used extensively in spacecraft and missiles, typically for pressure tanks, fuel tanks and high-speed vehicle skins. Titanium's high strength, low-density and high melting point is also the reason that it is used in numerous missiles.

While titanium is used in some key structural applications that would be hard to fill with other materials, the overwhelming use (95%) of titanium is as titanium oxide. By far the biggest use of titanium oxide is in paints, but it is also used as a pigment in paper, plastics and even candy and sunscreens. It is one of the whitest substances on Earth and has a high refractive index of 2.43. The iridescent colors of thin anodized layers of titanium oxide of different thicknesses is utilized in jewelry. It can also be added to surfaces to make them self cleaning.

The high strength and low density of titanium mean that the highest strength titanium alloys have a very high strength to weight ratio (~300 kN.m/kg) similar to that of the highest strength maraging steels. They also have good fatigue resistance (resistance to cyclic loading), but not very high ductility. The latter feature is related to its crystal structure. Many metals have a cubic crystal structure, whereas at room temperature titanium is hexagonal, see Fig. 45.2, which is a crystal structure associated with limited ductility because of the lack of easy deformation modes. Above 882 °C, the crystal structure changes to body-centered cubic, see Fig. 45.2. The two different forms are referred to as alpha (α) and beta (β) titanium.

The intermetallic compound titanium aluminide (TiAl) alloy (with a composition in at. % of Ti-48Al-2Nb-2Cr) was patented by Dr. S.C. Huang at General Electric in 1993 and was first introduced into gas turbine engines in 2011: the General Electric GENX – 1B engine was used in the Boeing 787 and the GENX-2B engine was used in the Boeing 747–8 with 400+ blades per aircraft. TiAl, see Fig. 45.3, has both lower density (4000 kg/m^3) and a higher specific strength than

Fig. 45.2 The crystal structure of titanium (top) at low temperature, which is hexagonal, and (bottom) the crystal structure above 882 °C, which is body-centered cubic

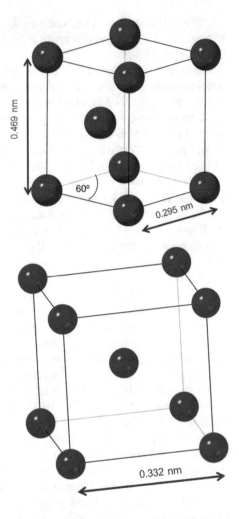

the titanium alloys previously used in the high-pressure compressor disks and blades and will undoubtedly spread to other gas turbine engines because of the weight savings it allows, leading to more fuel efficient engines. TiAl not only has superior specific strength compared to Ti alloys from 200 to 900 °C but also has a higher specific strength than steels and nickel-based superalloys, see Fig. 45.4. TiAl has also found applications in turbochargers in various vehicles because of its high strength at elevated temperatures, low density and corrosion resistance.

In 2016, World Titanium sponge production was 170,000 tonnes, with China, Japan, and Russia being the top three producers at 60,000 tonnes, 54,000 tones and 38,000 tonnes, respectively. The titanium market is growing by about 5% per year. As mentioned earlier, growth is not limited by the availability of titanium but by the

Fig. 45.3 The tetragonal
crystal structure of the
titanium aluminide TiAl

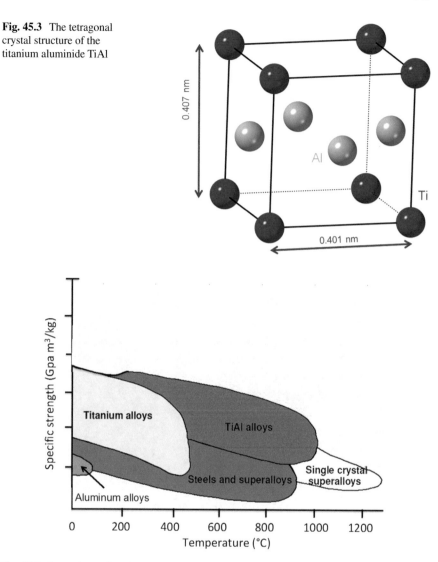

Fig. 45.4 Strength as a function of temperature for various materials

cost of processing. Thus, there are efforts underway to devise new cheaper processing routes such as that being pursued by Metalysys, a spin-off from Cambridge
University, located in Wath-upon-Dearne in South Yorkshire, England, which is
attempting to devise electrolytic methods to win titanium from its ore. Lower cost
titanium could find many more applications in petrochemical processing and offshore structures, including ocean-based wind turbines.

Chapter 46
Tungsten

Tungsten, a lustrous silvery-white metal, is one of the so-called refractory metals that includes tantalum, molybdenum, niobium and rhenium that all melt above 2475 °C – tungsten has the highest melting point amongst metals at 3422 °C. The refractory metals are hard and relatively chemically inert, but unfortunately generally have poor oxidation resistance. Tungsten's crust abundance is only 0.00011%. Tungsten, was discovered by the Spanish chemist brothers, Fausto de Elhuyar (1755–1833) and Juan José Elhuyar Lubize (1754–1796) in 1783. Apart from in French and English, tungsten is called Wolfram in most European languages, hence its Periodic Table symbol is W. The name comes from "tung sten" the Swedish for "heavy stone", which arises because it is one of the densest metals at 19,300 kg.m^{-3}. Tungsten is one of only seven elements with densities over 17,000 kg.m^{-3} and the only element that is of practical use in most density-driven applications, the others are two expensive, radioactive, or too chemically reactive.

Tungsten is extracted from the ores Wolframite, $(Fe,Mn)WO_4$, and Scheelite, $CaWO_4$. These are processed by various chemical methods to produce tungsten oxide, which can be heated under flowing hydrogen or reacted at elevated temperatures (1050 °C) with carbon to produce tungsten powder.

Tungsten components are usually made by compaction and sintering of powders because the very high melting point makes a melting and casting route commercially non-viable. While high-purity single crystal tungsten is ductile, generally polycrystalline tungsten has little ductility at room temperature. In high-purity form it can be worked by hot forging, hot drawing, or hot extrusion, but, as with many b.c.c. metals, impurities harden tungsten and also make it brittle. Because of the high temperatures needed for processing, tungsten is quite expensive.

Tungsten is very strong and very stiff with yield strength of up to 1200 MPa, an ultimate tensile strength or fracture strength (UTS) of around 1500 MPa and an elastic modulus (stiffness) of 405 GPa at room temperature.[1] By comparison, a 1090 mild steel has a yield strength of 530–620 MPa, a UTS of 800–970 MPa and an

---

[1] https://www.plansee.com/en/materials/tungsten.html

© Springer International Publishing AG, part of Springer Nature 2018
I. Baker, *Fifty Materials That Make the World*,
https://doi.org/10.1007/978-3-319-78766-4_46

Fig. 46.1 Scanning electron micrograph showing the microstructure of a typical tungsten heavy alloy, 93 W-5Ni- Fe (by wt. %), which was sintered at 1580 °C for 2 h. The large rounded grains are tungsten and the light regions consist of iron and nickel, which glue the tungsten grains together. (Courtesy of Randall German)

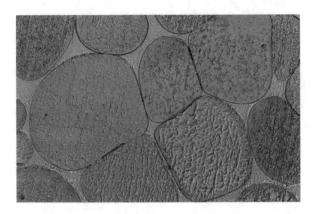

elastic modulus of 210 MPa.[2] Tungsten also has the best creep strength of any metal over 1650 °C.[3] It has the lowest thermal expansion coefficient of any pure metal at 4.5×10^{-6} °C^{-1}; by comparison, copper is 17×10^{-6} °C^{-1} and iron is 12×10^{-6} °C^{-1}.

So-called *Tungsten heavy alloys* are used in many applications. These contain 90–98% tungsten with a transition-metal binder containing a combination of nickel, iron, cobalt, manganese or copper.[4] They can be considered composites since they consist of tungsten grains in a transition-metal matrix, see Fig. 46.1. The matrix becomes liquid at around 1450 °C, which enables powders of the alloys to be sintered together at a much lower temperature than pure tungsten. However, tungsten heavy alloys are not usable at high temperatures like tungsten since the transition metal matrix softens and eventually melts. However, unlike tungsten, tungsten heavy alloys can be readily machined, and they have low reactivity, low toxicity and good mechanical properties. For example, for 91 W-Ni-Co can exhibit an ultimate tensile strength of up to 1700 MPa and up to 40% elongation, but not at the same time since there is a trade off between strength an elongation. They have been used to replace lead and uranium in a number of applications. Since they are 89–96% of the density of pure tungsten, they are used as balancing weights in a number of applications, for radiation shielding, and even as shot for shooting waterfowl where lead shot is banned.[5] Tungsten heavy alloys are also used as kinetic energy penetrators to pierce armor plating. They are replacing depleted uranium in this role since the latter's toxicity and slight radioactivity is problematic.

The tungsten-filament incandescent light bulb was the principal source of lighting for about 100 years, but it is now largely obsolete having been replaced by the compact fluorescent light bulb and more recently the light emitting diode (LED). Many countries are phasing out incandescent light bulbs because although they are cheap, 95% of the input energy turns not into light but into heat. In fact, the initial

[2] http://www.makeitfrom.com/material-properties/SAE-AISI-1090-1.1273-G10900-Carbon-Steel

[3] http://www.totalmateria.com/page.aspx?ID=CheckArticle&site=ktn&NM=110

[4] https://www.plansee.com/en/materials/tungsten-heavy-metal.html

[5] B-14-03925_KMT_technical_engineering_guide_Densalloy.pdf

Fig. 46.2 Tungsten filaments are used in X-ray tubes

cost of the light bulb is much smaller than the cost of the energy used over its lifetime. Although the use of tungsten filaments in incandescent light bulbs is disappearing, they are still used in x-ray tubes and electron microscopes. The heating of the tungsten filament not only produces light and heat it also "evaporates" electrons, which are used in those machines, see Fig. 46.2.

A Russian émigré to the United States Alexander de Lodyguine or Alexander Nikolayevich Lodygin (1847–1923) appears to have been the first to use tungsten as a light bulb filament and was awarded a U.S. Patent "Illuminant for Incandescent Lamps" in 1897 in which tungsten along with some other refractory metals were described. He also noted the well-known feature of tungsten - that it is hard and brittle, and, thus, he was unable to make a filament.[6, 7] That step has to wait until 1906 when William Coolidge (1873–1975) working at the U.S. company General Electric developed "ductile tungsten" through a combination of swaging and hot-drawing of sintered tungsten made from purified tungsten oxide.[8] The key was the swaging process, which involves repeated striking with hammers. This process produced a high-purity, fiber-like grain structure that could then be drawn.[9] Modern tungsten filaments typically contain potassium. At the operating temperature the potassium is present as liquid bubbles, which both reduce the creep (time dependent deformation) and prevent grain growth, which can be a detriment to the mechanical properties.

---

[6] http://russianheritagemuseum.com/en/RHM_Alexander_Lodygin/

[7] https://broom03.revolvy.com/topic/Alexander%20Lodygin&item_type=topic

[8] http://www.edisontechcenter.org/coolidge.html

[9] http://what-when-how.com/inventions/tungsten-filament-inventions/

Fig. 46.3 Tungsten
carbide balls

Tungsten is used in electric heaters and nozzles on rocket motors because of its high strength at high temperature and because it is a very good thermal conductor at 173 W/m.°C (copper is 400 W/m.°C and steel is 50 W/m.°C) and electrical conductor at 1.82×10^7 Siemens/m (copper is 5.8×10^7 Siemens/m and iron is 1.0×10^7 Siemens/m). However, more than 50% of tungsten is used in cutting tools and drill bits as tungsten carbide, which is extremely hard (9 on the Mohs hardness scale in which the hardest material, diamond, is 10), see Fig. 46.3. It is also added to some steels usually as the alloy ferro-tungsten, which is about 74–80% tungsten with the rest mostly iron. Tungsten's very close density to gold has led the nefarious to incorporate tungsten in gold bars, thereby replacing very expensive gold with much less expensive tungsten.

Tungsten was around \$35/kg in 2017.[10] Historically, the price has been quite volatile and has been as high as an inflation-adjusted price of \$107/kg in 1916. In 2016 China accounted for about 82% of world tungsten production of 86,000 tons with the second and third largest producers, Vietnam and Russia, accounting for only 7% and 3%, respectively. Tungsten is 0.007% of the Earth's crust and China accounts for half the reserves of tungsten ores.[11] Thus, China has a very strong influence on the market price. Much of the tungsten produced in China is used there. Tungsten production has been relatively flat in the last few years and around 30% of tungsten is recycled. The immediate future for tungsten is not that bright: as it is replaced in lighting applications in more and more countries, its usage may decline.[12] Use in lighting applications was 12% in 2008, which is declining by 5% per year.[13] However, tungsten use will continue to grow in tungsten-containing nickel-based superalloys, in tungsten steels (which contain 7% tungsten), in greater use of tungsten carbide and in mill products. Its use in microelectronics is also expanding since tungsten has a similar thermal conductivity and thermal expansion coefficient to silicon.[14]

---

[10] https://www.metalary.com/tungsten-price/

[11] https://minerals.usgs.gov/minerals/pubs/commodity/tungsten/mcs-2017-tungs.pdf

[12] http://www.prnewswire.com/news-releases/tungsten-market-set-to-undergo-further-changes-to-2018-262359771.html

[13] http://www.prnewswire.com/news-releases/tungsten-market-set-to-undergo-further-changes-to-2018-262359771.html

[14] http://www.strategyr.com/MarketResearch/Tungsten_Market_Trends.asp

Chapter 47
Uranium/Uranium Oxide

Uranium, a silvery white metal, was isolated in 1841 by French chemist Eugène-Melchior Péligot (1811–1890) [1] and named after the planet Uranus. Later, in 1896 French physicist Antoine Henri Becquerel (1852–1908) [2,3] showed that uranium salts and uranium were radioactive and, hence, discovered radioactivity for which he received the Nobel prize in 1903 sharing it with the French husband and wife team Marie Skłodowska-Curie (1867–1934) [4] and Pierre Curie (1859–1906) [5] who also made huge contributions to the study of radioactivity.[6,7] There are three naturally occurring isotopes of uranium (uranium has eight isotopes in all): uranium-238, uranium-235, and uranium-234, which have half-lives of 4.6 billion years, 700 million years and 25 million years, respectively. The long half-life of uranium-238, which is close to the age of the Earth, means that it is by far the most abundant isotope (99.27% of any natural uranium) with the short half-lived uranium 234, which is a decay product of uranium-238, constituting only 0.0050–0.0059% of natural uranium [1, 2].[8,9] Uranium constitutes only about 0.00018% of the Earth's crust.[10]

Uranium exists in three different crystals structures (see Fig. 47.1): up to 668 °C it is orthorhombic (alpha); from 668–775 °C it is tetragonal (beta); above 775 °C until it melts at 1132 °C it is body centered cubic (gamma).

---

[1] https://www.researchgate.net/publication/236233020_Eugene_Melchior_Peligot

[2] https://www.britannica.com/biography/Henri-Becquerel

[3] http://www.nobelprize.org/nobel_prizes/physics/laureates/1903/becquerel-bio.html

[4] https://www.chemheritage.org/historical-profile/marie-sklodowska-curie

[5] http://www.nobelprize.org/nobel_prizes/physics/laureates/1903/pierre-curie-bio.html

[6] https://www.britannica.com/science/uranium

[7] https://www.britannica.com/science/uranium#ref205381

[8] http://www.rsc.org/periodic-table/element/92/uranium

[9] http://education.jlab.org/itselemental/ele092.html

[10] http://periodictable.com/Properties/A/CrustAbundance.al.html

© Springer International Publishing AG, part of Springer Nature 2018
I. Baker, *Fifty Materials That Make the World*,
https://doi.org/10.1007/978-3-319-78766-4_47

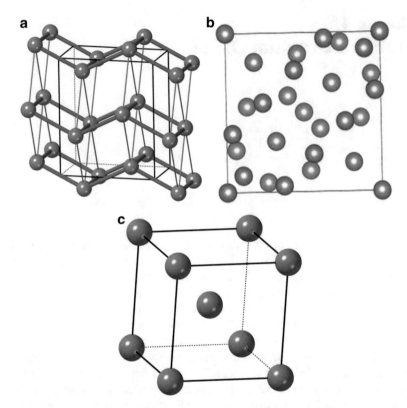

Fig. 47.1 Crystal structures of (**a**) orthorhombic alpha ($a = 0.285$ nm, $b = 0.587$, $c = 0.495$), (**b**) tetragonal beta ($a = 1.076$ nm, $b = 0.566$), and (**c**) cubic gamma ($a = 0.353$ nm) uranium. The alpha structure model shows more than a single unit cell to indicate the puckered layer structure of the atoms. The beta structure shows the complex arrangement of the atoms. The beta uranium crystal structure is from https://crystallography365.wordpress.com/2014/04/ and the VESTA. (Visualization for Electronic and STructural Analysis) web site and Momma and Izumi, "VESTA 3 for three-dimensional visualization of crystal, volumetric and morphology data," *Journal of Applied Crystallography*, 44, 1272–1276 (2011) are acknowledged

Uranium oxide, U_3O_8, is obtained mostly from the mineral Uraninite or Pitchblende that after various processing yields a bright yellow solid referred to as yellowcake. Uranium trioxide and uranium dioxide have long been used in glass and porcelain and as a glaze to provide different colors ranging from green, black and orange depending on the firing conditions.[11] Uranium dioxide is used in some nuclear reactor fuel rods either alone or mixed with plutonium oxide.[12] Like uranium both these oxides are only slightly radioactive. However, all are poisonous. U-235 is the fissionable isotope used in nuclear weapons and many nuclear power plants, but the concentration in yellowcake is far too low at 0.72% for this purpose.

---

[11] http://uranium.atomistry.com/uranium_trioxide.html

[12] https://www.britannica.com/topic/ceramic-composition-and-properties-103137#ref609025

Thus, the U-235 proportion has to be increased. This is done by turning the uranium oxide into gaseous uranium hexafluoride and then using high-speed centrifuges and diffusion membranes to separate the hexafluoride containing U-235 from the hexa-fluoride containing U-238, a process used in the Manhattan project. The result is a small amount of enriched uranium that contains 3–5% U-235, and a large amount of waste called depleted uranium (DU).

DU, which contains only 0.2–0.4% of the most radioactive uranium isotope U-235, although still slightly radioactive, has found a number of applications as load balancing weights in planes, medical radiation shielding and containers for radioactive materials because of its high density.

DU also has military applications for both armor plating and ordnance. DU has been used in some ammunition because of its density of 18,900 kg/m³, but it is largely being replaced by tungsten, which has a similar density of 19,600 kg/m³. The most well known use of DU is as a 30 mm incendiary round fired from the General Electric GAU-8/A Avenger Gatling-gun-type canon of the A-10 Warthog at 4200 rounds per minute.[13] The Warthog is famed as a tank killer. In the first and second Iraq wars around 320 tonnes and 900 tons of DU munitions were used, respectively.

DU is also utilized in Kinetic Energy penetrators that are used to kill tanks [3]. These consist of a long thin DU projectile in a discarding sabot that is fired from a howitzer or another tank. For most metals, when they hit a target, the metal smashes against the target and forms a mushroom shape. In contrast, uranium exhibits a self-sharpening action where parts shear off to make the projectile pointed upon hitting the target. A possible replacement for DU are single crystals of tungsten aligned along the cube axis of the crystal, which can also produce the same phenomenon. Unfortunately, these are very expensive to make. Another feature of DU projectiles is that they are flammable. Thus, when a DU projectile bursts through the wall of a tank it will be white hot and burn as it bounces around inside the tank.

Before uranium was used for nuclear power, it was used in the first nuclear weapon, named "Little Boy", which was dropped on Hiroshima, Japan in 1945.[14] Now, the major use of uranium is as fuel for nuclear reactors. It's future there is uncertain. Some countries, such as the U.K. and China, are building new reactors, whereas some countries, such as Germany and Japan, are ending nuclear power use. The huge costs and long time frame for building a nuclear power plant make it unattractive compared to natural gas and even renewal energy sources such as wind or solar in some countries, such as the U.S.A., where energy production is largely private.

In 2016, 62 metric tonnes of uranium were mined. Three countries mined 71% of the World's production: these were Kazakstan (39%), Canada (22%) and Australia (10%).[15] The price of uranium (oxide) has been falling with time from a spot price of around $97/kg at the beginning of 2013 to around $51/kg at the end of 2017.[16]

---

[13] http://www.military.com/equipment/gau-8-avenger

[14] http://www.atomicheritage.org/history/little-boy-and-fat-man

[15] http://www.world-nuclear.org/information-library/nuclear-fuel-cycle/mining-of-uranium/world-uranium-mining-production.aspx

[16] https://www.cameco.com/invest/markets/uranium-price

References

1. Chaline, E. (2012). *Fifty minerals that changed the course of history*. Buffalo: Firefly Books Ltd..
2. Stwertka, A. (2012). *A guide to the elements*. New York: Oxford University Press ISBN: 978-0-19-983252-1.
3. Danesi, M. E. (1990). *Kinetic energy penetrator long term strategy study*, The AMCCOM Task Group, Task Group Leader. (24 July, 1990), www.dtic.mil/get-tr-doc/pdf?AD=ADA395913.

Chapter 48
Wood

Along with stone, wood is the oldest material utilized by man. Originally it was indubitably used for fuel, for weapons and in construction of buildings and bridges, see Fig. 48.1. It was even used for keys by the Ancient Egyptians 4000 years ago [1].

Wood has featured prominently in all forms of transportation: on land in chariots, sleds, wagons and carriages; at sea as ships – most ships were made of wood until the end of the nineteenth century and many smaller boats are still made of wood; – and perhaps, most interesting in aircraft. Americans Orville (1871–1948) and Wilbur (1867–1912) built and flew the first powered aircraft, the aptly named Wright Flyer I, at Kill Devil Hills, North Carolina in 1903. It was built of spruce, which is both strong and lightweight, and the propellers were also made of wood. Most of the aircraft that immediately followed were also built from wood. Nevertheless, aluminum was used to make propeller blades as early as 1907 and the first mass-produced plane using aluminum construction, the Breguet 14, was introduced in 1916. The Breguet 14 was a single engine bomber and reconnaissance biplane, designed by Frenchman Louis Breguet (1880–1955), which was fast, agile and because it was able to sustain substantial damage ushered in the switch to metal aircraft construction.

However, even in the Second World War wood was still sometimes used for military aircraft construction. The most famous is the British twin-engined de Havilland Mosquito that was almost completely made of wood and when introduced in 1941 was one of the fastest aircraft in the world. It was produced up until 1950 and was used by the British, Canadian, Australian, and American air forces in roles as a light bomber, a fighter-bomber, a night fighter, a reconnaissance aircraft and a marine strike aircraft.

Apart from the Wright Flyer, probably the most famous wooden aircraft was the Hughes M-4 Hercules, nicknamed the Spruce Goose, although it was actually made of birch. The plane, which had eight engines, garnered a number of records: it had the largest wingspan of any aircraft; it is the largest aircraft made from wood; and it was the biggest flying boat ever built. Only one was built and this made only one

© Springer International Publishing AG, part of Springer Nature 2018
I. Baker, *Fifty Materials That Make the World*,
https://doi.org/10.1007/978-3-319-78766-4_48

Fig. 48.1 The bridge that traverses the Connecticut River connecting Cornish, New Hampshire and Windsor, Vermont is the longest wooden covered bridge at 137 m long (7.3 m wide) in the U.S.A. that carries both foot traffic and automobiles. (Courtesy of Alan Baker)

Fig. 48.2 Cross-section of woody cell showing several layers in cell wall [7]

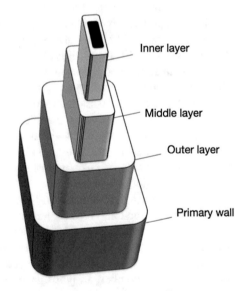

Inner layer

Middle layer

Outer layer

Primary wall

flight of 1.61 km (one mile) in 1947 with Howard Hughes (1905–1976), the designer and owner of Hughes Aircraft Company flying it.

Wood can be considered both to be a natural cellular composite material and a hierarchical structure (see Fig. 48.2) constituted from micro-fibrils of several organic polymers: cells made of cellulose (~40–50%), a polymer consisting of 5000–10,000 mers or repeat units run along the trunk or branch held together in a

Fig. 48.3 The overall structure of a tree trunk or branch

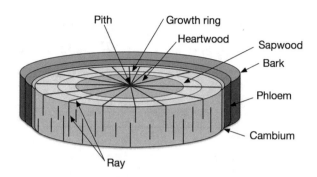

matrix of hemicellulose (which constitutes from ~20% in deciduous trees to ~30% in conifers) and lignin (which constitutes from ~23% in deciduous tress to ~27% in conifers) [1]. The lignin is a complex polymer that provides rigidity to the wood and also does not easily decompose.

A tree trunk or branch consists of many parts (see Fig. 48.3): the Pith is a small region right in the center of the trunk that is often pulpy; the Heartwood surrounds the pith and is the majority of the tree trunk in older trees, which although it can be considered dead, supports the tree and can be water reservoir for the surrounding Sapwood; the Sapwood, which encircles the Heartwood and also supports the tree, through which minerals and water flow up the tree; the thin Cambrium layer around the Sapwood that is a layer in which new cells form that add to both the Sapwood and the surrounding Phloem; the thin Phloem layer that transports products from the leaves to the rest of the tree; and the Bark, which protects the tree from drying out and from dangerous temperatures. In addition to these concentric features of the tree truck, the Rays both store and transport nutrients laterally through the trunk.

Figure 48.4 shows both a softwood such as a pine, fir or yew, and a hardwood such as oak or cherry. The names softwood and hardwood are somewhat misleading since some softwoods such as yew can be hard, while some hardwoods such as balsa can be soft. The key difference between hardwoods and softwoods is that hardwoods have a more complex structure containing large pores through which water is transported, the size of which varies greatly with the species of tree. The lignin in softwoods is largely derived from coniferyl alcohol, while that in hardwoods is derived from both coniferyl alcohol and sinapyl alcohol. Softwoods also show more marked differences in Earlywood produced early in a growth season and the denser Latewood produced later in the season, which contributes more to the strength. These are observed as annual growth rings, although some tropical hardwoods such as teak and mahogany may not show growth rings at all.

The very anisotropic microstructure of wood, that is the structure is different in different directions, results in different properties in different directions. For example, tensile strength values for Douglas Fir range from 2 MPa perpendicular to the grain structure to 88 MPa along the grains.[1] Plywood is way of using wood that

---

[1] http://worldwideflood.com/ark/design_calculations/wood_strength.htm

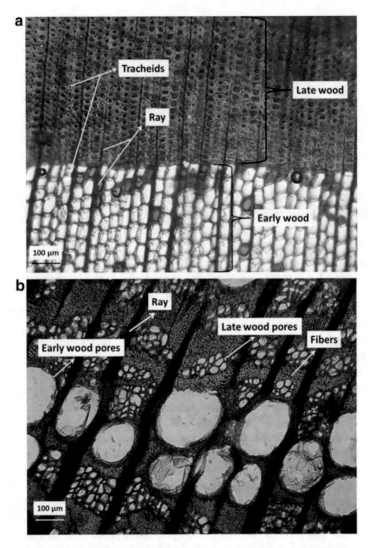

Fig. 48.4 Images of (**a**) a softwood Picea (coniferous evergreen spruce trees) and (**b**) a hardwood Ulmus (Elm). (Courtesy of Sara Essert)

avoids to some extent its anisotropic properties. It is made by gluing, using both heat and pressure, always an odd number of layers of wood in which the grain direction is at right angles in alternating layers. Cross-laminated timber, developed in Europe in the 1990s is a more advanced version of plywood in which two by four timber is glued along its edges to form large sheets, which are then pressed together, with the layers in adjacent layers running perpendicular to each other. Such sheets can span 18 m.[3] Other advanced forms include glue-laminated timber and parallel strand timber.

Turning to the mechanical properties of wood, the modulus of elasticity (resistance to stretching) for Douglas Fir is 13 GPa is much higher than a common polymer such as polyethylene (0.8 GPa), but is significantly less than a metal such as nickel (170 GPa). The modulus of elasticity varies considerably between different woods from 3.7 GPa for Balsa wood to 21 GPa for Black Ironwood. The Janka Hardness scale is used to compare the relative hardness of different types of wood. Values range from 22 for Cuipo and 100 for Balsa to 660 for Douglas Fir to 995 for Cherry to an amazing 5060 for the Australian Buloke. A wood such as Douglas fir is up to 3.5 times the strength of steel per unit weight [2].

The strength and elastic modulus of wood show clear relationships with its density, with the density of wood varying between species. The density of wood typically is 300–700 kg/m^3, that is always less dense than water (1000 kg/m^3), but Balsa wood is considerably less with values between 110–200 kg/m^3.

The strength of an actual piece of wood depends on knots, which are the remains of dead branches or dormant buds. Knots are imperfections that reduce the strength, particularly by increasing the ease of splitting along the wood grain, but do necessarily affect the modulus. The strength of wood also depends on the water content. Below the fiber saturation point (typically 25–30% moisture), in which the cell walls are still completely saturated but no water remains in the cell cavities, the strength of wood depends strongly on the moisture content, whereas above the fiber saturation point the strength of wood is independent of the moisture content. Dry wood can be up to four times the strength of green, that is, undried wood [1].

Drying wood results in shrinkage and since the structure of wood is anisotropic the shrinkage is different in different directions. Dry wood has thermal expansion coefficients of 3–5 × 10^{-6}/°C and 20–60 × 10^{-6}/°C parallel and perpendicular to the grains, respectively, this is a factor of 5–15 different in the thermal expansion coefficients in the different directions [3]. The thermal conductivity of wood is low at (0.12–0.04 W/m.°C) compared to metals such as aluminum (205 W/m.°C) and even less than ice (1.6 W/m.°C) and some polymers such as polyethylene (0.33–0.5 W/m.°C), but comparable to Polyurethane (0.02 W/m.°C).

Hardwoods are employed in a large range of applications, including tools, construction, boat building, furniture making, musical instruments and flooring, but 80% of the wood used is softwood, which is mostly used for construction. Some woods are used for very specific applications. Baseball bats are mostly made of ash, while cricket bats are made of white willow, which is lightweight (dried density of 400 kg/m^3), tough and is not easily dented. Many sports originally used wooden equipment, but have now changed to other materials. For instance, while wooden (nowadays usually plywood) sticks are still used in ice hockey most are now carbon fiber composites. Similarly, some musical instruments are made of specific types of wood because of their specific acoustic properties. For example, ripple sycamore is used for the backs of violins, while African blackwood is used for clarinets, and European spruce is used for violin soundboards [1].

Tally sticks made of wood, ivory, bone or stone have been used for thousands of years to record data. The split tally stick made of wood was used in Europe in the Middle Ages, most prominently in England. Henry I of England, who was also the

Duke of Normandy, instituted the split tally stick system as part of the Exchequer's system for collecting taxes in around 1100 A.D., a system that was continued until 1826. In the English split tally system the details of a loan were made on a willow, pine or hazel-wood stick. The stick was then split into two parts of unequal length. The lender was given the longer part of the stick called the Stock and the shorter part, called the foil, resided with the debtor, possibly giving rise to the expression "getting the short end of the stick". The tally sticks thus became a form of money. When the tally stick system was finally abandoned in 1826, the sticks were stored in the Houses of parliament until 1834. In that year, the unfortunate decision was made to burn the remaining two cartloads of tallies in a furnace in the House of Lords. The furnace used was a coal furnace and burning the sticks led to a fire that burned the House of Lords and eventually spread throughout parliament burning down the House of Commons. The construction of the replacement Palace of Westminster started in 1840 [4, 5].

Wood is likely to have a bright future. It is, of course, biodegradable, and it is a way of capturing carbon. The carbon footprint from building a wooden structure can be quite favorable compared to other materials. There is also flurry of building tall timber structures. While tall structures, such as the 32 m Temple of the Flourishing Law in Nara Prefecture, Japan built 1400 years old ago, have been constructed in the past, a new era is emerging of building tall timber structures such as the 49 m high 14-storey Treet apartment block that was recently completed in Bergen, Norway [2] and the 53 m high, 18-storey Brock Commons student accommodation at the University of British Columbia, Vancouver, Canada, which was completed in 2017 [3], while the 73 m HAUT residential tower will be started in Amsterdam, The Netherlands that year.[4] Such buildings, which are produced of laminated wood, sometimes with concrete layers for strength and fire protection, can be built more quickly, quietly and with less construction traffic than steel and concrete structures [6]. Wood also requires less energy to produce than steel or concrete.

References

1. High-security locks. *The Economist*, April 15th, 2017.
2. Tall Timber. (2016). Science, 23rd Sept, Vol 353 pp. 1354–1356.
3. Flinn, R. A., & Trojan, P. K. (1990). *Engineering materials and their applications* (4th ed.). Boston: Houghton Mifflin Company ISBN: 0-395-43305-3.
4. Unusual Historicals/Money Matters/The Tally Stick System.pdf.
5. The Tally Stick BBC podcast: 50 Things that made the Modern Economy, May 26th 2017.
6. Top of the tree. *The Economist*, Sept 10th, 2016, pp.66–67.
7. Foulger AN. (1969). *Classroom demonstrations of wood properties*, U.S. Department of Agriculture, Forest Service, Forest Products Library PA-900.

---

[2] http://www.timberdesignandtechnology.com/treet-the-tallest-timber-framed-building-in-the-world/

[3] https://www.skyscrapercenter.com/building/brock-commons-phase-1/22424

[4] http://www.archdaily.com/791703/the-netherlands-tallest-timber-tower-to-be-built-in-amsterdam

Chapter 49
Wool

Domestication of sheep started at least 10,000 years B.C. Sheep not only provided wool, but also milk and meat as well. Spinning of wool also started around this time. Before wool was processed into fabric, undoubtedly, it was worn as a woolly skin taken from a dead sheep.

Wool was the source of England's prosperity in the Middle Ages, and English wool was the most prized in Europe. This was recognized in the English Parliament where in the fourteenth century the Lord Speaker of the House of Lords was commanded to sit on a large wool-stuffed cushion with a small backrest called the Woolsack, a tradition that still continues.[1] While England was the source of wool, the production of cloth from this wool drove the prosperity of the Low Countries, France and Italy. Eventually, wool exports from England fell as the wool was mostly turned into cloth for export and by the mid-fifteenth century more broadcloth was exported than wool.

The central nature of wool in England means that wool and its technology added words to the English vocabulary. For example, the word "tease" is derived from the use of the dried flower head of Fuller's Teasel, which is a natural three-dimensional comb that is used to "tease", that is, raise the nap on wool and other fabrics. Later, hand Carders were used for this purpose, see Fig. 49.1. Teasels grow wild in England and as a child I recall picking them and dyeing the heads for decoration. Presumably the expression "woolly thinking" arises because sheep are not perceived as the smartest of animals.

While England's prosperity was based on wool from sheep, wool can be produced from many animals such as llamas, goats (cashmere), muskoxen (qiviut), rabbits (angora) and alpacas, whose wool can be much more valuable than the wool from sheep.

Sheep have both hair, referred to as kemp, and wool and the wild sheep first domesticated had more hair than wool, a feature that selective breeding has inverted. The first evidence of such selective breeding is from Iran around 6000 B.C. with the

---

[1] http://www.woolsack.org/woolsackhistory

© Springer International Publishing AG, part of Springer Nature 2018
I. Baker, *Fifty Materials That Make the World*,
https://doi.org/10.1007/978-3-319-78766-4_49

Fig. 49.1 "Hand Carders" that are used to disentangle, clean and mix wool fibers. (Woodstock History Center, Woodstock, Vermont, U.S.A.)

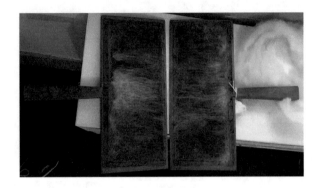

earliest wool clothing dating from to 3000–4000 B.C.[2] Wool has a protective coating composed of hygroscopic (that is absorbs water from the air) protein (keratin) fibers created in the sheep's skin. The fibers themselves are dead. Wool is distinguished from fur in that it grows in clusters or staples. Also, unlike fur, which grows to a fixed length and stops growing, wool grows continuously. This continuous growth is a trait shared with the hair on sheep. In fact, some sheep, typically those bred in warmer climates, grow more hair than wool. Hair, which is also a protein fiber composed largely of keratin, doesn't have to be sheared like wool. Like wool, it grows continuously, but it is shed every spring.

Sheep wool fibers are typically 12–50 microns (a micron is one thousandth of a meter) in diameter, 4–150 mm long from the finest to coarsest fibers with overlapping scaly cuticle cells on their surface, which are about 10% of the fiber (Figs. 49.2 and 49.3). A waxy coating on the cuticle cells makes them water repellant, but they can still absorb water vapor. Because of the protective waxy coating, wool fibers are resistant to staining. During processing of wool these scales interlock together enabling dense fabrics to be produced. Wool fibers are generally white but can also be brown, black, silver or a combination of these. The fibers are wavy with around 2–12 waves per centimeter from coarsest to finest fibers, which is a measure referred to as the crimp. The crimp defines how well the fibers can be spun into cloth. A fine crimp, which is usually composed of fine fibers having many waves, has the capacity to be spun into fine cloth, which has greater market value. The keratin fibers consist of an amorphous matrix (about 42% of the volume) along with crystalline filaments. The latter are insoluble in water, but the former can absorb water and expand [1]. There are two different types of cells in the fiber, orthocortical and paracortical. The difference between these two types of fibers is that the paracortex has a higher proportion of both matrix and the amino acid cysteine than the orthocortex. The two types of fibers expand by different amounts when they absorb water and this difference in behavior leads to the crimp (Fig. 49.2). Finer wool fibers (less than around 25 microns) are typically used for clothing, whereas coarse fibers are used for rugs, coats and jackets.

Wool has many useful properties. One feature that may not immediately come to mind is that wool absorbs sound. Woolen clothes have greater bulk than other fabrics, which also allows it to entrap air, a feature, which along with its high specific

---

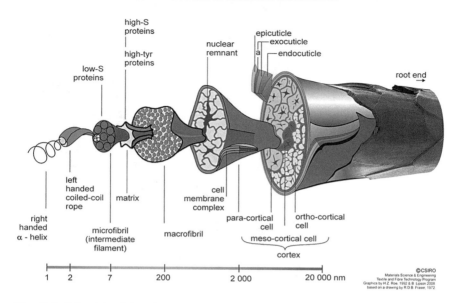

Fig. 49.2 Schematic of the microstructure of a wool fiber (http://www.scienceimage.csiro.au/ image/7663//large)

heat of 1.3 kJ/kg.°C (similar to cotton), prevents heat transfer and so it is an excellent insulating material for both warm and cold climates. It can absorb water up to 30% of its own weight. As wool absorbs water vapor, it becomes heavier and warmer to wear. Since it is hydrophilic it can easily absorb and retain dyes (dyeing is typically performed at >60 °C in a slightly acidic water). However, the dyeing process slightly damages the wool with higher temperatures producing more damage [2]. While wool decomposes at 130 °C, it is also naturally flame retardant because of its high nitrogen and water content. It also has a high ignition temperature and lower heat of combustion than man-made fibers, will not melt and will not continue to burn once a flame is removed. Thus, it is often used in firefighter's and soldier's garments. Wool's low density (1300 kg/m³) – a slightly higher density than water (1000 kg.m⁻³) - allows lightweight fabrics to be produced. Wool, being a natural fiber, is biodegradable, and because of its high nitrogen content will typically biodegrade in about a year, a significant advantage over synthetic fibers.

Wool fibers are quite strong. If wool is pulled, it exhibits three sequences of behavior as shown schematically in Fig. 49.4[2]. Initially, a linear elastic region occurs up to yielding at 2% elongation. This yield point is associated with a change in the crystal structure of the keratin in the crystalline filaments. Unloading at any point below 2% strain allows the wool to return to its original length. From about 2–30% elongation the so-called yield region occurs where the stress increases more slowly than in the elastic region as the wool is pulled. Interestingly, if this wool is unloaded in this region and soaked in warm water it will again completely recover its original length. Beyond about 30% elongation the stress required to pull the wool further increases more rapidly until fracture occurs. Figure 49.4 also shows how much of

Fig. 49.3 Scanning electron micrographs of two different types of wool. The wool was coated with gold so that it would not electrically charge in the scanning electron microscope. (Courtesy of Zhangwei Wang)

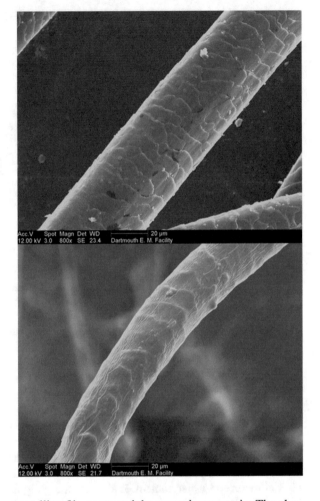

the stress is carried by the crystalline filaments and the amorphous matrix. The elastic modulus (stiffness) of wool is 2.5–4.5 GPa depending on the type sheep from which the wool is taken, which is greater than that of most synthetic polymers. The strength and elongation depend on the relative humidity, the strength decreasing with increasing humidity and the elongation increasing. At 0% humidity, wool can show a yield strength of 150 MPa (compare to high-purity aluminum at 7–11 MPa), while at 100% humidity it can show 60% elongation upon failure.

Wool subjected to sunlight, specifically UV light, tends to lose it strength and yellows[4]. In fact, wool shows an exponential decrease in strength when subjected to light of UV wavelengths less than 300 nm if it is at temperatures greater than 35 °C (temperature alone has little effect) [3] with the strength falling to less than 20% of its original value at 45 °C. This is important in applications such as upholstery in cars, but the effect is mitigated since the windows in cars filter out the UV wavelengths.

Fig. 49.4 Schematic of a stress–elongation curve for a wool fiber, showing the stress accrued by the crystalline filaments and the amorphous matrix

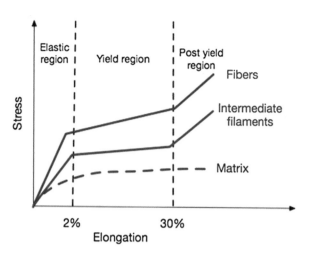

While no other natural or synthetic material can match the outstanding qualities of wool, wool is a small fraction of the fabric produced in the world because of its high cost (5–6 times that of other fibers). Over the last 25 (35) years, both the absolute volume of wool has fallen and as a percentage of all textile fiber production it has decreased from 4.3% to 1.3% [3]. Currently, around 50% of the world's wool of around 1.1 million tonnes is produced in just three countries China, Australia, and New Zealand [4] with China consuming about 60% of the world's production. Interestingly, wool production [5] in the U.S.A. has fallen from 50,000 tons in 1980 to around 15,000 tons in 2015.[6] Wool will likely continue to be used as a high-quality fiber for clothing and rugs into the future, but is unlikely to regain its dominance as a clothing material. This has led to a trend towards producing finer wool fibers for clothing that is next to the skin[4].

References

1. McKittrick, J., Chen, P.-Y., Bodde, S. G., Yang, W., Novitskaya, E. E., & Meyers, M. A. (2012). The structure, functions, and mechanical properties of keratin. *Journal of Metals, 64*, 449–468.
2. Huson, M. G. (2009). Tensile failure of wool. In A. R. Bunsell (Ed.), *Handbook of tensile properties of textiles and technical fibres* (pp. 100–143). Cambridge: Woodhead Publishing ISBN 978-1-84569-387-9.
3. Holt, L. A., & Milligan, B. (1984). Evaluation of the effects of temperature and UV-absorber treatments on the photodegradation of wool. *Journal of Textile Research, 54*, 521–526.

[3] http://blog.mecardo.com.au/wool-market-share-is-it-really-important-or-just-a-distraction

[4] International Wool Textile Organization, wool statistics 2015.

[5] http://www.agmrc.org/commodities-products/livestock/lamb/wool-profile/

[6] https://www.nass.usda.gov/Charts_and_Maps/Sheep_and_Lambs/w_shorn.php

Chapter 50
Zinc

You have most likely encountered zinc as a dietary supplement. Even though the pill container may say "Zinc", it isn't zinc metal, it's likely zinc gluconate, zinc acetate or zinc oxide, see Fig. 50.1. This medicinal use of zinc accounts for little of the actual use of zinc. Zinc is a hexagonal-close-packed (h.c.p.), see Fig. 50.2, bluish-white metal that is relatively hard, and has low ductility. It is reactive and tarnishes in air forming a corrosion-resistant oxide. Zinc is the fourth most commonly-used metal after iron, aluminum and copper even though it is only the 24th most common element at 0.0078% of the Earth's crust. It is typically found in ores containing lead and copper with the most common ore being mined as a form of zinc sulfide called Sphalerite. Thirty-eight percent of the 2016 World mining production of 12 million tonnes was extracted in China with the next largest producer, Peru, generating less than 30% of China's production.[1] Metallic zinc is produced by turning zinc sulfide into zinc oxide by roasting at elevated temperature, which is then turned into zinc by either a pyrometallurgical processing using carbon or an electrolytic processing. Roughly, 30% of zinc is produced by recycling.

Even though zinc was known by the Romans, it wasn't until twelfth century India that it was produced in significant quantities, and it was not used significantly in Europe until the sixteenth century. However, brass, a gold-colored copper-zinc alloy, has been used since the third millennium B.C. starting in the Middle East. Brasses have a wide variety of applications and are often used for decorative applications because of their gold-like appearance, see Fig. 50.3. The most common alpha brasses or yellow brasses, which are 65–70 wt. % copper and 30–35 wt. % zinc, consist of zinc atoms dissolved in a copper lattice to form a substitutional solid solution. Annealed yellow brass is not very strong with a room temperature yield strength of only around 95 MPa although it shows very high ductility at 65% elongation and good fracture strength of over 300 MPa. However, if cold-worked the yield strength can be increased to over 400 MPa with a concomitant reduction in elongation to less than 10% [1].

---

[1] mcs-2017-zinc.pdf

© Springer International Publishing AG, part of Springer Nature 2018
I. Baker, *Fifty Materials That Make the World*,
https://doi.org/10.1007/978-3-319-78766-4_50

Fig. 50.1 The container although labeled "Zinc", really contains Zinc Gluconate, which is used as a dietary zinc supplement

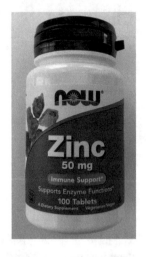

Fig. 50.2 The hexagonal close-packed structure of zinc. The ratio of the axes at 1.86, is far from the ideal packing of a hard spheres in a hexagonal array of 1.63

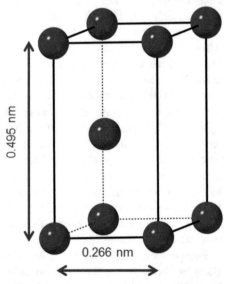

Brass is used for radiator cores for automobiles, ammunition cases, zippers, low-friction uses such as gears, bearings, locks, for plumbing and electrical uses, and for applications where flammable materials or explosives are present since it will not spark when struck. It is also used extensively for musical instruments, horns and bells. Probably the most famous brass bells are 13.8-tonne "Big Ben", the nickname of the Great Bell that chimes out the hours in the Elizabeth Tower (originally St. Stephen's tower) at the Palace of Westminster, London,[2] which was possibly named after a 1.93 m high heavyweight pugilist Benjamin Caunt who had that nickname,[3]

---

[2] http://www.visitlondon.com/things-to-do/sightseeing/london-attraction/big-ben#mUHtgc 10XQu8JV6Y.97

[3] Secrets of Westminster, PBS, aired 2014-09-08.

Fig. 50.3 A typical use of   switch cover
brass as an electrical

and the 900 kg Liberty Bell in Philadelphia, PA. Both these bells were cast by the Whitechapel Bell Foundry Ltd. in London, which was founded in 1570 and is the U.K.'s oldest manufacturing company.[4] It is notable that both of these bells cracked soon after first use; Big Ben was subsequently repaired, while the Liberty Bell was recast by local workers but again cracked in the early nineteenth century.[5]

There are, in fact, numerous types of modern brasses some of which can have up to 67 wt. % zinc and some of which contain aluminum, arsenic, manganese, silicon and/or phosphorus. Brass can also have a different meaning: 'where there's muck there's brass' is a twentieth century expression that originated in Yorkshire, England where brass is sometimes used as a slang term for money.

About 55% of zinc is used for galvanization in which steel is dipped in molten zinc to coat it and prevent rusting. Such galvanized steel is used in car bodies, bridges, and for corrugated steel or iron sheets for building low-cost housing. The zinc does not just coat the iron or steel to prevent it rusting, it has a galvanic action. In other words, if some of the zinc is scratched off the steel an electrical current between the zinc and steel acts to protect the steel with the zinc acting as a sacrificial anode preferentially corroding before the steel. Zinc is often used in block, plate, rod or even a ribbon form as a sacrificial anode for ship hulls, propellers and rudders on small boats, on offshore platforms and pipelines in saltwater environments.[6]

Zinc is sometimes used for roofing. Paris has its very distinctive look since 85% of roofs are made of rolled zinc sheet,[7] a feature that has its origins in the large public works program overseen by Baron Georges-Eugene Haussmann (1809–1891) during the reign of Napoleon III (1808 –1873).[8] Such zinc roofs have a lifetime of 80–100 years.

---

[4] http://www.whitechapelbellfoundry.co.uk/history.htm

[5] http://www.ushistory.org/libertybell/

[6] http://www.azom.com/article.aspx?ArticleID=749

[7] https://wadearch.com/zinc-roofing/

[8] https://www.britannica.com/biography/Georges-Eugene-Baron-Haussmann

Fig. 50.4 A zinc die-cast
toy car

As for the rest, about 16% of zinc is used in brasses and bronzes, 21% in other alloys, with the balance used to make many zinc compounds, particularly zinc oxide and zinc sulfide.[9] It is added to rubber, plastics and paints. A zinc disk coated with more expensive and more aesthetically-pleasing copper has been the basis of the U.S. one cent coin since 1982.

Because zinc is relatively inexpensive, has a low melting point of 420 °C and does not react with steel dies, zinc alloys are used for die-casting. Such alloys, which typically contain 4% aluminum, 0.05% magnesium and sometimes around 1% of copper. They have good room temperature tensile strengths of around 400 MPa[10] with relatively low elongations of 7–10% [1] – the strengths are superior to those of some of the stronger precipitation-hardening aluminum alloys and of alloys of another h.c.p. metal magnesium. These alloys also have good corrosion resistance and, because of the high thermal conductivity and low viscosity of molten zinc, castings can be made inexpensively with high dimensional stability that need little secondary machining. Die-cast zinc alloys are easily painted, plated polished, anodized and are used extensively for automobile parts, hardware, electrical components, and even toys, see Fig. 50.4.

A small, but historically important use of zinc is in batteries. The first voltaic pile was made from alternating layers of zinc and copper separated by felt or cardboard soaked in an electrolyte of sulfuric acid or brine by the Italian physicist and chemist Alessandro Volta (1745–1927) in 1799. Batteries were developed over the years until the German physician and scientist Dr. Carl Gassner (1855–1942) produced the first dry battery in 1886 using a zinc cup as the negative electrode. The modern dry battery, which produces 1.5 V, is the zinc-carbon battery, which is an evolution of the wet Leclanché cell invented by the French scientist Georges Leclanché

---

[9] http://metalpedia.asianmetal.com/metal/zinc/application.shtml

[10] http://www.accucastinc.com/zinc_die-casting.html

(1839–1882) in 1886 that uses an ammonium chloride electrolyte, a manganese oxide depolarizer, a carbon rod cathode that runs the length of the battery and the tubular zinc case as the anode.

Zinc's use is likely to increase at about the same rate as the World economy increases with galvanized steel for buildings and cars as the main use. Zinc is one of the most recycled materials and can be recycled indefinitely.

The price of zinc bounces around, but it has risen considerably from $1529/tonne at the beginning of 2013 to $2700 at the end of 2017. By comparison, copper is nearly $7000/tonne and aluminum is around $2180/tonne.

References

1. Flinn, R. A., & Trojan, P. K. (1990). *Engineering alloys and their applications*. Boston: Houghton Mifflin Company ISBN: 0-395-43305-3.

Printed in the United States
By Bookmasters